T0318358

Solutions to Environmental Problems Involving Nanotechnology and Enzyme Technology

Solutions to Environmental Problems Involving Nanotechnology and Enzyme Technology

Alka Dwevedi

Academic Press is an imprint of Elsevier
125 London Wall, London EC2Y 5AS, United Kingdom
525 B Street, Suite 1650, San Diego, CA 92101, United States
50 Hampshire Street, 5th Floor, Cambridge, MA 02139, United States
The Boulevard, Langford Lane, Kidlington, Oxford OX5 1GB, United Kingdom

Library of Congress Cataloging-in-Publication Data
A catalog record for this book is available from the Library of Congress

British Library Cataloguing-in-Publication Data
A catalogue record for this book is available from the British Library

ISBN 978-0-12-813123-7

For information on all Academic Press publications
visit our website at https://www.elsevier.com/books-and-journals

Publisher: Candice Janco
Acquisition Editor: Candice Janco
Editorial Project Manager: Emily Thomson
Production Project Manager: Sojan P. Pazhayattil
Cover Designer: Greg Harris

Typeset by SPi Global, India

Contents

Preface

A coin is incomplete without its two sides, and the same applies to any released innovative technology: one must look at all sides for its complete long-term success

Chapter 1: Nanotechnology has become a crucial factor in almost every sector (industry, agriculture, water treatment, energy storage, electronics, etc.). It appeared in 1959 in the form of manipulating matter at the atomic and molecular levels; however, it began to really boom from the beginning of the 21st century. Solutions for current energy demand and environmental remediation problems are the two main areas addressed in this book. The exponential increase in global population is a cause of greater energy demand and the need for environmental remediation, due to increasing living standards. It has been estimated that energy usage will increase by about 40% in the coming 20 years and will double by 2050. Energy demand is directly related to economic development, climate change (responsible for air and water pollution), global health, national security and even to poverty.

Presently, we are dependent on fossil fuels like coal and petroleum, either directly or indirectly, for most of our energy requirements. However, these sources are responsible for generating toxic gases in addition to greenhouse gases and they create other serious problems, most of them having no solutions until very recently. We have come to a state where solutions are a must, to save our future. Renewable energy sources like solar, wind, and water are the best alternatives as they are freely available and generate no toxic by-products. We need appropriate technologies that can effectively harness their energy and provide it to us for our varied applications. Solar energy can be easily harnessed through photovoltaics, converting light energy into electrical energy. However, its initial installation cost is quite high and is usually beyond the reach of ordinary people. Several nanomaterials have been found that can be effectively used in photovoltaics, reducing its cost by several hundred folds in addition to providing a large increase in efficiency. Further, nanomaterials have led to miniaturization of photovoltaics and they can now easily fit into computer cases, mobile electronic devices, transporting vehicles, smart coatings for glass, and even into clothing (for conversion of sunlight into electrical energy). Nanomaterials can be used in producing artificial photosynthesis and generating H_2 (clean energy sources with more caloric value than any fuels known so far). Windmill blades can be easily modified using nanomaterials to enhance their efficiency in generating electricity. Even storage batteries can be made more efficient as well as nontoxic by incorporating nanomaterials.

Nanomaterials can also be helpful in many other ways: protecting available energy sources by detecting even microscopic leaks in oil pipelines; providing better catalytic efficiency in fuel production from raw petroleum; producing better fuel consumption in vehicles and power plants by increasing combustion efficiency and decreasing friction; reducing resistance generated in high tension wires used in electric grids, among other applications. Nanomaterial coatings are being used on vehicles such as cars, trucks, airplanes, boats, and even spacecraft, to lower fuel usage by several folds.

Global industrialization has been increasing at an exponential rate, thus severely affecting our natural resources. Maintenance and restoration of air, water, and soil quality have become major challenges for our age. Various remediation technologies are available, but none of them are very efficacious in removing all the present toxic contaminants. Nanotechnology has been useful in providing solutions by remediating various toxic contaminants within a short time span, due to their very large surface area-to-volume ratio, effective quantum confinements, and catalytic properties. Various nanosorbants, nanocatalysts, bioactive nanomaterials, and nanostructured catalytic membranes are known to have excellent efficiency in removal of various toxic contaminants in air and water (many of these are discussed in this book). Water-repellant nanomaterials are also providing solutions to oil spillage problems, which cause major havoc to aquatic ecosystems. Nanomaterial-based air filters are being fitted in airplane cabins for cleaning released harmful particles and odors via nanofilters. Nanomaterial-based sensors have been developed for monitoring treated air and water quality, at even sub-ppm levels.

Nanotechnology is undoubtedly contributing to many solutions for energy and environmental problems, and is a powerful tool for environmental sustainability. However, there has been heated debate over its toxicological effects on ecosystems and human health. Several issues are concerning. Most nanomaterials are not being experimentally evaluated for their toxicity. Further, there are issues related to release of toxic by-products responsible for environmental hazards during synthesis of nanomaterials. The size of nanomaterials is responsible for their easy mobility into living cells and induction of toxicological effects. Most importantly, high costs are associated with running pilot and field trials to investigate nanomaterials. Additionally, excess amounts of energy and water are expended during nanomaterial fabrication. It has been estimated that an average of 15 years would be required for thorough analysis and validation of nanomaterial risk assessments. All of these issues have been major obstructions of release of nanomaterial-based products into commercial markets.

There is an urgent need for additional technology to mitigate any adverse effects of nanotechnology. Enzyme technology, based on enzyme immobilization onto matrices, has been effective in this area, due to the lowering of the surface energy of nanomaterials, imparting additional catalytic properties, increasing specificity, and resisting aggregation, a major complication when a nanomaterial enters a living system. Further, enzyme technology has been helpful in addressing the energy crisis and environmental remediation under mild physico-chemical conditions in presence of non-toxic traditional chemicals. Nanomaterial-based immobilized enzymes are easily separable from the reaction mixture, making them completely free from the nanomaterials (responsible for any toxicological effects) and thus nullifying the exposure toward the nanomaterials. This combination of enzyme technology and nanotechnology has economical benefits, as both technologies have large market potential around the world. Enzymes immobilized onto nanomaterials have excellent physicochemical properties as compared to any other known matrices for enzyme immobilization. It has been reported that enzyme-based nanomaterials are excellent sensors for detection of various contaminants present in our surrounding air, water, or soil.

Thus, the combination of the two technologies has been helpful in solving various problems, with additional advantages due to favorable economical and environmental conditions. In this chapter, a summary of research related to nanotechnology for the last several decades has been compiled, along with a discussion of toxicological effects of nanomaterials and the need for combining nanotechnology with enzyme technology.

Chapter 2: Energy is the basic requirement for our survival, as it has become an integral part of our daily life. Energy requirements are increasing cumulatively due to our enhanced living standards and increased industrialization. Energy is the integral part of every sector of our lives; our industries, transportation, the myriad comforts and conveniences of home and workplace, and even the security of our nations are all derived from various sources and forms of energy. A number of reports have been published based on the linkage between economic growth and energy consumption. The available statistically significant data indicate that energy consumption is related to economic growth, represented by gross domestic product (GDP) per capita, with an 0.82% increase in GDP in upper middle-income countries, 0.81% increase in GDP in lower middle-income countries, and only a 0.73% increase in GDP in lower-income countries when energy consumption has increased by 1% in all cases. The appetite for energy is endless, but the sources of energy are circumscribed (~85% of total energy is derived from fossil fuels like oil, coal, and gas). The finite store of fossil fuels and emission of various toxic gases, which represent major health threats in addition to causing global warming and ozone depletion, have further propelled our search for other effective substitutes. Various renewable energy sources available on earth are long lasting, clean with no generation of toxic gases, cheap, and easily accessible. They can provide long-term availability of energy supply, increase diversity of energy sources, promote regional development (can be used even in undeveloped areas without conventional energy sources), and effectively reduce the cost associated with climate change. A technology is needed that can effectively capture energy stored in these renewable sources in a cost-effective manner. This chapter deals with the utilization of nanotechnology, as well as the combination of nanotechnology and enzyme technology, in harnessing energy from renewable sources in cost-effective ways.

Chapter 3: The availability of safe water has been a growing problem, and it is becoming worse with increasing urbanization and population density. Only about 2% of fresh water can be used for human purposes. It has been estimated that over 3.5 billion people will be in a water scarcity condition by 2025, based on the current population growth rate. Advanced water treatment technology has been helpful in dealing with the water scarcity problem but this technology is largely available only to developed countries. The problem has become more exacerbated due to introduction of various recalcitrant, nondegradable compounds from agricultural and industrial activities. Even developed countries are not able to cope with these, as there is no suitable technology available that can efficiently remove all types of recalcitrant compounds. Children (0–8 years) are at major risk from consumption of contaminated water leading to various neurological diseases, weakening of the immune system, and arrested growth. Over 1.8 million (4.1% of total global deaths due to various diseases) human deaths have been reported by WHO (World Health Organization) annually due to consumption of contaminated water. Nanotechnology has provided cost-effective and efficacious solutions for these water treatment issues by removal of all types of recalcitrant compounds in addition to microbial load, including viruses. Besides using nanomaterials, like carbon nanotubes (CNTs), nanosorbents, dendrimers, etc., for water treatment, they can also be used for water desalination, disinfection, and in sensors that can sense contaminants even at sub-ppm concentrations. They can even remove toxic heavy metals like arsenic, organic materials, salinity, nitrates, pesticides, etc. from surface water, groundwater, and wastewater. Most importantly, nanomaterials can carry out water treatment without any addition of chlorine (known to be controversial due to generation of carcinogenic compounds).

However, due to its potential toxicological risks for humans and the environment, nanotechnology has not gained much ground in the commercial market for water treatment. The combination of enzymes and nanomaterials has been shown to lessen the toxicological impact of nanomaterials while retaining their water purification, disinfection, sensing and monitoring abilities, with excellent specificity and catalytic efficiency. Further, this combo-technology has been effective in removal of biofilm (which usually blocks nanofilters) by degrading its components containing polysaccharides and proteins. Most importantly, the combination technology can reduce cost by several folds due to increased stability and cycles of reusability. This chapter outlines water purification, disinfection, sensing, and monitoring using the combination of the two technologies.

Chapter 4: Polluted air has become a major health hazard worldwide, with developing countries being the topmost target, because they lack updated technologies for air pollution checks and control. WHO in collaboration with the University of Bath (United Kingdom) has collected worldwide data on air quality and found that there has been an increase of over 6% of air pollutants per year and this will increase to higher levels if it remains unchecked. The major factors behind this increase in air pollution are geographic and atmospheric conditions, scale and composition of economic activity, population, strength of local pollution regulations, and the energy mix. There are over 3 million deaths (about 11.6% global deaths) every year due to air pollution associated with both indoor and outdoor air quality. WHO has given guideline limits for an annual mean of $PM_{2.5}$ at $10\,\mu g\,m^{-3}$ (includes air pollutants such as sulfate, nitrates, mineral dust, ammonia, and black carbon), as they are related to many health risks, particularly lung and cardiovascular system disorders; *tissue and systemic inflammation; increased oxidative damage to DNA and cell membrane lipids; increased risk for thrombosis, lowered birth weight and impairment of metabolic, cognitive, and immune function;* developmental delays in children; premature death; reproductive health problems, etc. Even developed countries are also at risk due to air pollution; for example, Europe has suffered from high levels of ammonia and methane gases generated from diesel-powered cars and farming policies, and the United States has been affected by ozone, etc. Therefore, an urgent need exists to control air pollution around the world using economical and efficient sustainable technology.

Nanotechnology has been helpful in controlling air pollution in several ways. For example, nanocatalysts (including nanofiber of manganese oxide) are being used to convert toxic gases into harmless gases much faster, due to their large surface area. Nanostructured membranes with predefined pore size are used to trap harmful gases. The NANOGLOWA project has developed a range of nanostructured membranes to replace scrubbers and significantly reduce CO_2 in emissions. GE has developed nanostructured membranes that can effectively remove CO_2 so it can be used as fuel. Carbon nanotubes (CNTs) have gained much significance lately, due to their several hundred times higher capacity for trapping harmful gases as well as their ability to detect them, making CNTs the most desirable technology for air purification for industrial-scale plants. Enzyme immobilization onto various nanomaterials has been found to enhance the air-purifying capabilities and sensing of toxic air pollutants by several thousand folds. Various successful commercial platforms have been efficiently operating on the concept of combined enzyme and nanotechnology, such as NanoTwin Technologies Inc., Green Envirotec, etc., for air purification. This chapter covers reported immobilized enzymes onto nanoparticles for air purification and detection.

Chapter 5: The combo-technology of nanomaterials and enzymes has provided solutions for many problems of the energy crisis and environmental remediation. However, acceptance of any technology depends not only on its efficacy and cost, but on various other factors to be considered. For the combo-technology discussed in this chapter, these factors include sustainability, energy consumption, and the promotion of ecological revivification by restoring natural flora and fauna. Further, the social and economical issues need to be properly dealt with in a transparent manner towards the public as well as governments, before these products are allowed to be released into the market. The combo-technology discussed in this chapter needs to fulfill these requirements before enjoying real success on the global platform. Combo-technology should contribute toward sustainable development (environmental, social, and economic) while following the path of human well-being in addition to economic support, with aspirations for a clean and healthy environment and contributions toward social development. It would be interesting to study the effect of combo-technology on the long-term development opportunities in poorer countries, along with its role in environmental remediation and energy production, in addition to its easier adaptability with the present technology being used for environmental remediation and energy production in various parts of the world. Richer countries can easily experiment and take risks in adopting combo-technology, due to their stable economic growth. However, in the case of developing and underdeveloped countries, long-term sustainability is a must for its adoption to take place. Further, new technology adoption also requires adequate skills, knowledge, and competencies, all of which should be carefully addressed.

The tests faced by any new technology, and its long-term survival, are based on its sustainability parameters. Further, nature itself will impose its own restrictions when a new technology is not fit or sustainable.

Updating technology has a direct correlation with the GDP of a country. However, a strong regulatory framework for environmental protection needs to be in place before release of any new technology. Further, the social effect of any new technology is another important parameter contributing to long-term success of the technology, as it defines directly the distribution of income, both within and across nations, as well as the reduction of poverty. Coordination between national and international policies, in addition to a sharp focus on technology after-effects on human and animal health, plant life, and conservation of natural resources, should be carried out for the ultimate success of an introduced technology. The globalization and sustainable development of new technology depends on simple technology transfer, substantial tariff reductions, and support from international financial institutions and multilateral development banks. Further, there should be disclosure of all the facts, including any side effects, and transparency with respect to social and environmental accountability, employment opportunities, and environmental consequences after release of technology in the international marketplace. This chapter gives an overview of the sustainability of combo-technology and its efficacy in the domestic and international spheres, besides addressing its environmental, social, and economic consequences.

This book is the outcome of combined efforts from various disciplines to understand the concepts and present them in an easier way. The main theme behind this book is to view technological advancements from various angles, rather than presenting only allure and speculation. Special thanks to Prof. Arvind M. Kayastha (School of Biotechnology, Faculty of Science, Banaras Hindu University) and Dr. Dinesh P. Singh (Department of Physics, University of Santiago) in giving their precious time in contributing to this book. I am thankful to Dr. Sandhya Dwevedi (Institute of Advanced Research) for helping me in various sections requiring physics expertise. I am thankful to Raman, Divya, Yogesh, Mummy, and Papa for their continuous encouragement and for sparing their time for me to complete this book.

Alka Dwevedi

CHAPTER

Overview of combo-technology (nanotechnology and enzyme technology) and its updates

1

Alka Dwevedi

Department of Biotechnology, Delhi Technological University, New Delhi, India

Swami Shraddhanand College, University of Delhi, New Delhi, India

1.1 Introduction

Nanotechnology has gained significant importance in almost every sector, including industry, agriculture, water treatment, energy storage, electronics, etc. Although the field of nanotechnology appeared for the first time in 1959, for manipulating matter at the atomic and molecular levels, its wider applications began to flourish at the start of the 21st century. Fig. 1.1 shows an overview of the synthesis of nanomaterials and their usage across the world for the period 2017–24. Most significantly, their crucial role in environmental sustainability has been a major focus of researchers and academicians.

The global population is growing at a tremendous pace, and environmental conditions have worsened due to higher living standards, leading to inevitable increased energy requirements. It has been estimated that energy usage will increase by about 40% over the coming 20 years and it will double by 2050. Energy demand is directly related to economic development, climate change (responsible for air and water pollution), global health, national security, and even to poverty. Countries like the United States, with well-established mature economies, are also facing the issue of negative correlation of growth in energy consumption with respect to energy production. Coal and petroleum have been the most important sources of energy across the world. However, their limited sources and the cumulative nature of CO_2 emissions in the atmosphere have forced the search for alternative energy sources that are abundant and that have minimal environmental impacts. The most important alternatives are renewable sources like sun, wind, and water. Trapping energy from these sources in a sustainable and economical manner without generating any unwanted by-products is a real challenge for the scientific community. In this context, nanotechnology has provided a ray of hope, with claims to providing sustainable eco-friendly solutions.

As an example, solar panels are known to be a cost-effective source of electricity, but their initial cost of installation is quite high and it takes at least 5–8 years to recoup the investment. Nanomaterials have now been used to produce lower-cost photovoltaics converting solar energy into electrical energy in a more efficient manner.

Solutions to Environmental Problems Involving Nanotechnology and Enzyme Technology
https://doi.org/10.1016/B978-0-12-813123-7.00001-3

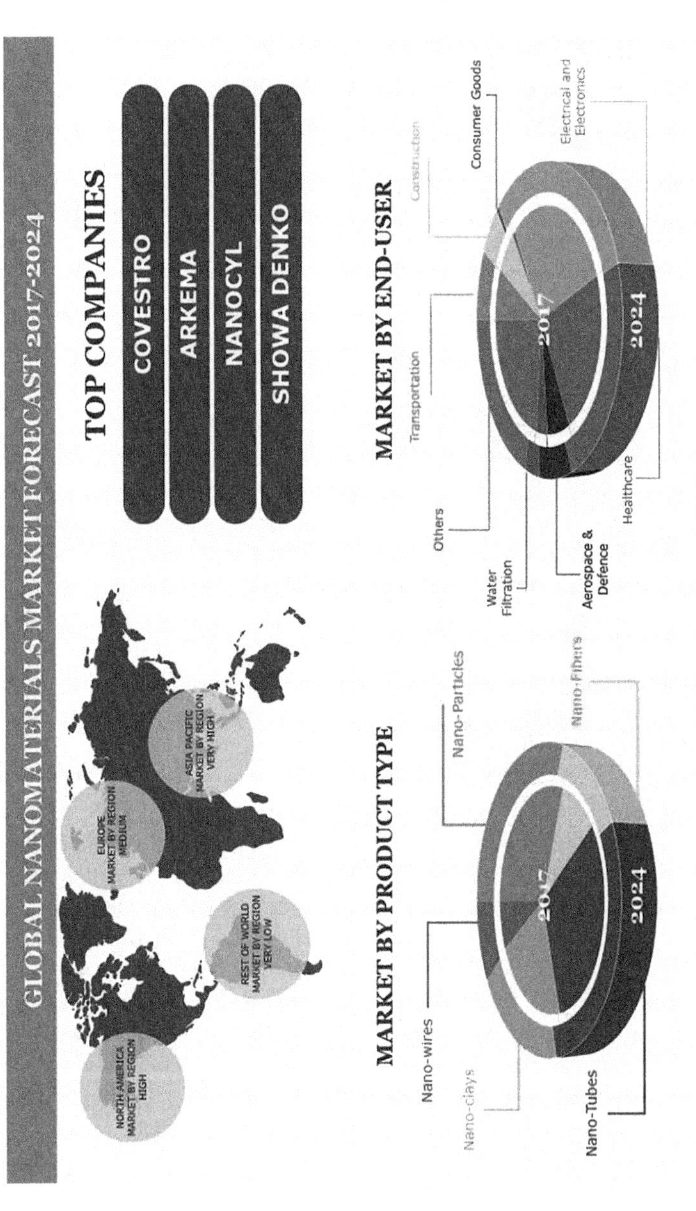

FIG. 1.1

Overview on global market for nanomaterials in the period 2017–24. The top four companies actively involved in the synthesis of nanomaterials are: COVESTRO, ARKEMA, NANOCYL, SHOWA DENKO.

Adapted from https://www.inkwoodresearch.com/reports/global-nanomaterials-market-forecast/.

These nanomaterial-based solar panels can be fitted onto computer cases, mobile electronic devices, transporting vehicles, smart coatings for glass, and even into clothing for economical conversion of sunlight into electrical energy. Further, solar energy can be used to help lower the level of CO_2 (major greenhouse gas) by using the gas in artificial photosynthesis in the presence of nanomaterial-based catalysts [1].

Windmills are also used in the generation of electricity, but their true efficiency and reliability have come into question. Now, various nanomaterials are being used in the manufacture of windmill blades to attain better length and strength and lighter weight, in order to improve the generating efficiency.

Nanomaterial-based batteries have been found to be a safer, cheaper, more efficient, more flexible, lighter weight, and more reliable alternative than other hazardous heavy metal-based batteries. These batteries offer better conduction, longer shelf life, faster recharging, and can easily be fused into any known electronics.

Nanomaterials are found to be helpful in detecting even microscopic leakages in oil pipelines and are used in fuel production from raw petroleum, providing higher catalytic efficiency than any other known methods. They are helpful in reducing fuel consumption in vehicles and power plants due to higher efficiency in combustion and decreased friction.

Carbon nanotubes (CNTs) are found to have less resistance than any other known high-tension wire used in electric grids, thus helping to reduce power loss during transmission. Nanomaterial coatings on vehicles such as cars, trucks, airplanes, boats, and even spacecraft have lowered fuel usage by severalfold. A NASA analysis of usage of nanomaterials in spacecraft found that there is a reduction of almost 63% of fuel consumption during vehicle launch as compared to conventional composites. Further, nanomaterials have also helped in increasing mission reliability and reducing launch costs by several hundredfold [2].

Recently, cellulose nanomaterials have been produced that are found to have wide applications in electronics, construction, packaging, heath care, automotive, food, energy, and defense. They are very cost effective with no side effects and are easily available. These are generally prepared from wood chips, cornstalks, grasses, etc. [3].

Global industrialization has progressed at an exponential rate, which has seriously affected our natural resources. Maintenance and restoration of air, water, and soil quality have been a major challenge in the present day. Water and soil in particular have been affected at a much faster pace, leading to deleterious impacts on all living organisms on earth. Almost all the countries of the world face environmental problems such as availability of drinking water, treatment of waste and wastewater, increasing problems with air pollution, and contamination of soil and groundwater. Various remediation technologies have appeared over the years, but none of them has been found to be efficacious in removing all the present toxic contaminants.

Nanotechnology claims to provide solutions by remediating all present toxic contaminants within a short time span by using a variety of nanomaterials having very large surface area to volume ratio, as well as effective quantum confinements and catalytic properties. These claims have attracted a worldwide market of around $23.6 billion; particularly the United Kingdom and Japan have seen a major expansion with respect to other parts of the world [4]. Nanotechnology has the ability to remediate and also to detect pollution and microbial inactivation at molecular levels, making it more profitable. Various nanosorbants, nanocatalysts, bioactive nanomaterials, and nanostructured catalytic membranes are known to have enhanced filtering ability, and thus are able to remove all contaminants, including toxic metal ions, radionuclides, and organic and inorganic solutes/anions present in water. Further, they can also be used for energy efficient water desalination.

However, there is a major challenge in treatment of industrial wastes particularly rich in chlorinated compounds. Nano-zerovalent iron (nZVI) has gained tremendous significance in this area due to its ability to remediate soil and groundwater, and it has been commercialized in several developed countries. In addition to remediating soil and water, nanotechnology can be used for purification of air containing various toxic gases like CO, VOCs, NOx, etc. [5]. Magnetic water repellant nanomaterials have been developed recently that can absorb oil about 20 times their weight, thus helping in oil spillages [6]. Many nanomaterial-based air filters are being fitted in airplane cabins to clean any released harmful particles using mechanical filtration. These filters have an additional charcoal layer that removes odors from the released air. These air filters can be used for various other applications [7]. Further, nanomaterial-based sensors have also been developed to monitor released air quality, even around fires. Therefore, nanotechnology can be a powerful tool for environmental sustainability and has the ability to supersede the conventional higher-cost technologies [8].

The effectiveness of nanotechnology in providing solutions for various environmental problems is not in doubt and it is a powerful tool for environmental sustainability. However, some controversy and debate have arisen regarding nanotechnology's toxicological effects on the whole ecosystem on being released into the environment. Most of the nanotechnology claims related to environmental sustainability are based on laboratory-scale experimentation without any verification from field usage. Only about 1% of the total nanotechnological applications for environmental use have been commercialized. Synthesis of nanomaterials involves drastic physical and chemical conditions that result in release of toxic by-products, posing additional environmental hazards. While nanomaterials are advantageous in degrading various recalcitrant environmental pollutants, additional risks exist of their easily mobilizing inside living cells and inducing a toxic response.

Nanomaterial development, including running pilot and field trials, is costly. An additional disadvantage is the expenditure of large amounts of energy and water during nanomaterial fabrication. Further, there are various technical challenges, the main one being the delivery of the particles to the target area. The known nanomaterials have incomplete ecotoxicological profiles, with little to no information about their potential health and environmental concerns. It has been estimated that an average period of 15 years is required for proper analysis and validation of nanomaterial risk assessments. There is no regulatory strategy at the present time for addressing the risk issues of nanotechnology and defining protocols for testing before their release into commercial markets [9].

Enzyme technology has been known for more than 50 years and is being thoroughly researched by scientists, academicians, and even in industry for its wide applications. More than 6000 enzymes are known, categorized into six classes (oxidoreductases, transferases, hydrolases, lyases, isomerases, and ligases) based on the type of reactions being catalyzed. They are easily available due to the presence of wide ranging sources, from prokaryotes to eukaryotes, with easier isolation due to recent advancements in chromatographic techniques. Further, they are highly specific, with several thousand times higher catalytic turnover than any other known alternatives under mild conditions in presence of traditional chemicals. They are easily biodegradable with no toxicity to the environment. LCA (life cycle assessment) and EIA (environmental impact assessment) reports on enzyme-based processes have strongly recommended their usage for wide applications in almost every known field of study. Its usage is also being emphasized due to its potential global market, which will be around 2.6 billion USD in 2024. North America is the largest consumer, followed by Western Europe. In the Asia/Pacific region, the highest enzyme demand is in China, followed by Japan and India. Further, economists have directly correlated increases in urbanization and GDP (gross domestic product) to the growing enzyme market. Therefore enzyme-based solutions to various problems are both economically and environmentally favorable. In the present context, enzymes are the wisest alternatives for production of clean, green energy and a solution for remediation processes being implemented for various pollutants in water, air, or soil [10, 11].

Enzyme-based energy production involves very cheap raw materials like lactose, sucrose, cellulose, xylan, steam-exploded aspen wood, starch, etc. The energy production is not only cost effective but has no greenhouse side effects such as those observed for fossil fuel–based energy sources. Mostly redox enzymes are used, particularly hydrogenases producing hydrogen as a source of energy. Hydrogen has been a powerful energy source with excellent combustive ability, with production of water during its combustion, which is not possible with any other known energy source. Hydrogenases can work only under anaerobic conditions with requisite metallic ions like iron and nickel. They are a very large protein with their active site containing an H-cluster being buried within their core. A fuel cell based on hydrogenase has been made to produce H_2 on an industrial scale with enzyme-based reduction at the cathode with the mediation of small electroactive molecules [12]. Further, degradation of lignocellulose biomass using cellulases has also produced clean energy via ethanol production as a fuel source. Several variants of cellulases with higher catalytic efficiencies are being explored to make the process very cost effective and easily adaptable by smaller-scale industries. Two variants in 2009 and 2010 were released by Novozymes, with the commercial name of Cellic followed by Cellic CTec2, and subsequently Cellic CTec3 was released in February 2012. The last version is 10 times better than previous versions, producing $2.00/gal ethanol and gasoline using a cellulosic source. Global market on enzyme-based ethanol production will reach USD 115.65 billion by 2025, with an increment of 6.7% CAGR (compound annual growth rate), leading to a market worth of ~$1.5 billion in 2015. It has been estimated that cellulosic ethanol will generate 15% of energy consumption by 2020 [13].

Enzymes are found to be very useful in transforming highly toxic pollutants coming from air, water, or soil into less toxic or innocuous products. Recalcitrant pollutants coming from different sources are polycyclic aromatic hydrocarbons (PAHs), petroleum hydrocarbons, phenols, polychlorinated biphenyls, azo dyes, organophosphorus pesticides, heavy metals, etc. Enzymatic treatments can be carried out with minimal additional chemical requirements under mild physical conditions without any environmental risks. Phenolic contaminants and related compounds coming from petroleum refining, metal coating, coal treatment, synthesis of resins and plastics, wood preservation, treatment of textiles, dyes and other chemicals, mining and dressing, treatment of pulp and paper, etc. can be easily treated using peroxidases (horseradish peroxidase, chloroperoxidase, lignin peroxidase, and manganese peroxidase). Recalcitrant chemical compounds (herbicides, pesticides, and insecticides) used for crop protection against herbs, weeds, insects, and fungal pathogens are a major challenge to the environment due to their nondegradation by various remediation techniques. Enzyme-based remediation has given much hope for their detoxification. For example, the enzyme parathion hydrolase (also known as phosphotriesterase) has been quite effective in degradation of organophosphate pesticides. Various other detoxifying enzymes are being explored for effective degradation of recalcitrant herbicides, pesticides, and insecticides.

Cyanide (extremely toxic to living organisms) is produced by a number of industrial processes like production of synthetic fibers, rubber, chemical intermediates and pharmaceuticals, ore leaching, coal processing, and metal plating. Further, it is also present in the effluents coming from industries based on food and feed production due to the presence of cyanogenic glucosides in various crop materials. It has been found that over 3 million tons of cyanide is being discharged yearly, in drinking water throughout the world. Two important enzymes, cyanidase and cyanide hydratase, can effectively convert cyanide into less toxic forms like ammonia, formate, or formamide. Pulp and paper waste rich in recalcitrant dyes, chlorophenol, and various bleaching agents can be easily treated with enzymes like laccase, cellulase, beta-glucosidase, etc. Heavy metals like arsenic, copper, cadmium, uranium, lead, chromium, strontium, etc. present in industrial wastes coming from electroplating, fuel reprocessing and pigment synthesis can be treated with phosphatase through precipitation followed by separation through different size screens.

Enzymes can be used for the detection and monitoring of various harmful pollutants present in air, water, or soil. Important enzymes are nitrate reductase, urease, laccase, tyrosinase, parathion hydrolase, cytochrome oxidase, acetyl cholinesterase, and peroxidase [10].

Enzymes are a versatile and very effective tool for solving problems related to the energy crisis and degradation of recalcitrant pollutants. However, there is an issue with their operational stability that needs to be resolved, to make them usable under all types of environmental conditions, even in the presence of hostile physical and chemical conditions. Various solutions have been suggested, the foremost being isolating enzymes from extremophiles, such as thermophiles, halophiles, acidophiles, etc., but it is not always possible to obtain the specific enzymes for an exact purpose from available extremophiles. Recombinant DNA technology provides the answer of modifying enzymes to make them resistant to hostile chemical and physical conditions. This technology has solved the problem to some extent but not completely, due to inefficiency of the host undergoing posttranslational modifications as is found in native enzymes, and improper protein folding, due to a variety of reasons. The most effective, simplest, and least laborious technique arrived at to solve the problem of enzyme stability in extreme conditions without affecting the native properties is enzyme immobilization. Most importantly, this technique allows easier enzyme separation after use in various applications, and also reuse depending on the reusability cycle of the immobilized enzyme. The physicochemical properties of an immobilized enzyme depend on the type of matrix used. Various matrices are known and their evolution continues in order to find a matrix that can impart maximum operational stability and selectivity, along with the highest immobilization efficiency and reduced inhibition of its enzyme products or any other components present in the reaction medium [10].

Nanomaterials have been used for enzyme immobilization and some very surprising results have been reported. Immobilized enzymes are highly stable in dry conditions, with very high reusability and catalytic turnover due to the very high surface-to-volume ratio of nanomaterials [10, 14]. This chapter provides an overview of the role of nanotechnology and enzyme technology in providing solutions to various environmental problems, as reported by current researchers. It also discusses the necessity of combining these two technologies and providing an appropriate platform for their wide acceptance and rapid commercialization.

1.2 Nanotechnology: research updates from the last 20 years

Nanotechnology has played a significantly important role in the protection of the environment and climate by saving raw materials, energy, and water as well as by reduction of greenhouse gases and hazardous wastes. This technology has provided solutions to various environmental problems, including climate change, need for clean drinking water, and environmental pollution. It has been estimated that the world's energy demand will double by 2050. With increasing population, a larger amount of raw materials will be needed for energy production, leading to havoc in the coming years due to limited resource supplies. Nanoproducts and processes have provided some solutions for energy production and storage. Some interestingly shaped nanomaterials with unique structures having applications in clean energy production and environmental remediation are shown in Fig. 1.2.

Most importantly, nanotechnology has helped to find solutions for clean and improved energy by utilizing solar, wind, hydrogen, batteries, and supercapacitors. Environmental pollution has been a very hot topic in every country of the world and it has led to destruction of our precious biodiversity and degradation of human health. It has been found that increased development with improved technological advances

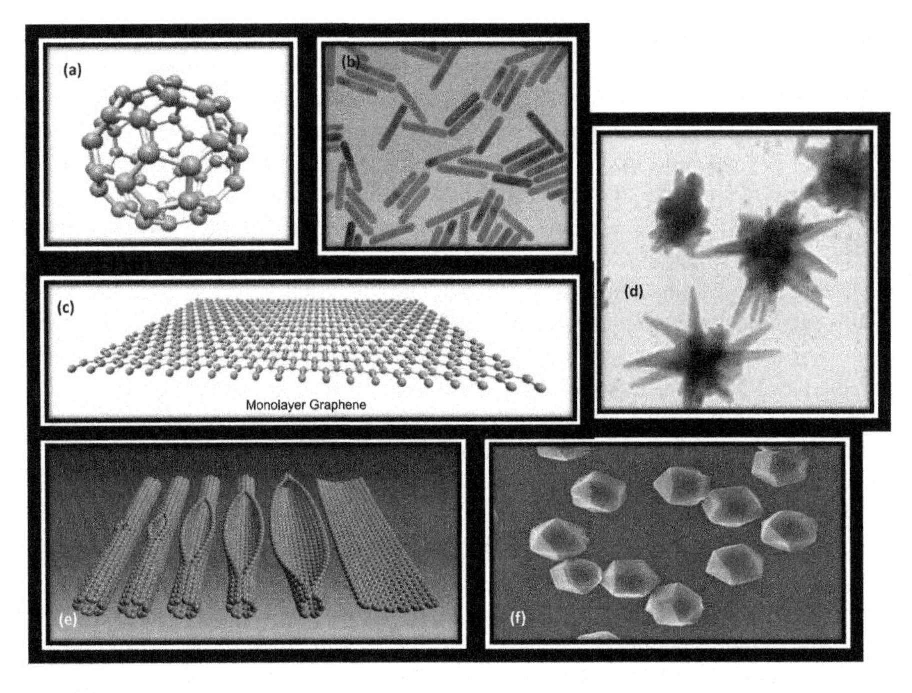

FIG. 1.2

Structures of some beautiful looking nanomaterials with wide applications. (a) Buckyballs (fullerenes) *(Courtesy: Lloyd-Hughes J, Jeon T-I. A review of the terahertz conductivity of bulk and nano-materials. J Infrared Millim Terahertz Waves 2012; 33: 871-925).* (b) Nanorods (Adapted from: https://nanocomposix.com/products/980-nm-resonant-gold-nanorods) (c) Monolayer (graphene) *(Courtesy: Lloyd-Hughes J, Jeon T-I. A review of the terahertz conductivity of bulk and nano-materials. J Infrared Millim Terahertz Waves 2012; 33: 871-925)* (d) Nanostars *(Courtesy: Tuan Vo-Dinh, Duke University)* (e) Nanoribbons *(Adapted from: http://compositesmanufacturingmagazine.com/2015/08/new-graphene-discovery-could-improve-future-electronics/)* (f) Nanodiamonds *(Adapted from: http://web.archive.org/web/19961117090637/http:/k2.scl.cwru.edu:80/cse/eche/faculty/angus/ico2dmd.jpg)*

have adversely affected our environment, for various reasons. Nanotechnology has found solutions by providing remediation and purification of water and air, as well as detection of contaminants and prevention of pollution. Fig. 1.3 provides an overview of publications on nanotechnology related to environmental remediation and production of clean energy for the last 20 years (data based on Pubmed). The following section discusses nanotechnology advances in providing green and clean energy with control over environmental pollution, over the last 20 years.

1.2.1 Clean and green energy

1.2.1.1 2001–2005

A highly efficient thin film electrode for solar cells was fabricated using nanowires of TiO_2. The network of TiO_2 nanowires was prepared in such a way that there was an

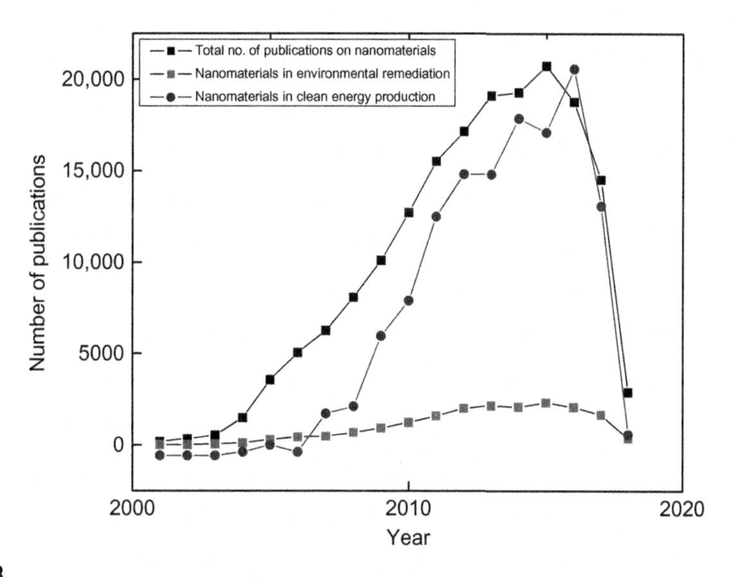

FIG. 1.3

Publications based on nanotechnology from 2000 to 2018, including applications of nanomaterials in environmental remediation and production of clean energy.

The data are collected from PubMed.

ordered alignment of nanowires leading to a single crystal like anatase, having excellent rate of electron transfer. Crystalline anatase had [101] plane with good efficiency of light adsorption leading to conversion yield of light to electricity of 9.3% [15]. The efficiency of a dye sensitized solar cell was improved by using crystalline ZnO nanowires as the anode. This set-up enhanced the surface area by several hundredfold, which improved the light-capturing capacity of the dye and thus enhanced the electricity generation by solar cells by several higher numbers [16]. The efficiency of the photovoltaic device was enhanced by several orders of magnitude using nanocrystalline TiO_2 and SnO_2 films. It was found that charge transport and recombination dynamics was 100-fold higher in the case of SnO_2 with respect to TiO_2, due to higher electron density. However, in both cases, the efficiencies of photovoltaic solar cells (dye sensitized) were improved due to faster regeneration of dye from its ground state [17].

Hydrogen and methane are known to be clean sources of energy, because burning of these gases does not release any toxic compounds. Further, they have more gross calorific value (H_2 : 39.4 GCV, CH_4: 15.4 GCV) than any other energy sources. A number of devices are being explored for adsorption of these gases for their storage. Carbon nanotubes (CNTs) have been found to be efficient carriers, as revealed from adsorption isotherms of H_2 and CH_4. The isotherm model for dry samples of H_2 and CH_4 behaved similarly, while the sorption behavior of CH_4 in a water loaded wet sample was quite different, having five times higher uptake capacity due to formation of methane hydrates. CNTs are known to be the best energy carrier due to their porous nature and higher surface area [18]. Efficient storage of H_2 has been explored for fuel cell applications by the US Department of Energy (DOE). Besides CNTs, nanographite platelets (graphenes) have been found to be quite efficient in their adsorption capacity. Graphenes have various advantages, such as light weight, low cost, chemical inertness, and environmental benignity [19].

1.2.1.2 2006–2010

There is an increasing demand for primary and rechargeable batteries powering various applications, from electronics to vehicles, without generation of much heat. They need to be lightweight, inexpensive, have higher energy density, and be eco-friendly. Tremendous efforts have been made to improve existing battery technologies based on new advanced electrochemistry, surface chemistry, and material chemistry. The idea behind these innovations has been to develop efficient energy storage devices. Nanomaterials have gained in significance due to their efficiency and advantages over bulk material in battery performance. Nanostructured active materials like Mg, Al, Si, Zn, MnO_2, CuV_2O_6, $LiNi_{0.8}Co_{0.2}O_2$, $LiFePO_4$, Fe_2O_3, Co_3O_4, TiS_2, and $Ni(OH)_2$ have found extensive applications in batteries. These nanostructured materials have numerous active sites with eloquent ionic transfer and diffusion, leading to higher battery efficiency with large capacity, high energy, and power density, as well as long cycle life. Further, due to their small size, they have improved kinetics and enhanced electrical conductivities [20]. The new generation lithium rechargeable batteries have gained in significance, being a source of clean energy with excellent safety, reliability, and cycling life. They are most commonly used in electronics, hybrid electric vehicles, and aerospace. Nanosized metals like $LiCoO_2$, $LiFePO_4$, and $LiMn_2O_4$ are used as cathodes in lithium batteries, leading to enhanced cycle life and efficacies [21].

Graphite is known for its anisotropic two-dimensional lattice. It is the most stable form of carbon under ambient conditions. It forms zero- and one-dimensional structures called fullerenes and nanotubes, respectively, in its nanometer range. Further, fullerenes and nanotube-like structures can also be prepared from numerous other inorganic compounds with almost similar characteristics. These types of structural nanomaterials are helpful in various applications like automotive and aerospace industries, home appliances, medical technology, etc. Further, they have potential applications in catalysis, rechargeable batteries, drug delivery, solar cells, and electronics [22]. Self-assembled monolayers of porphyrins and fullerenes act as donor and acceptor, respectively, when linked to gold electrodes in photochemical devices. These self-assembled monolayers are very effective in light harvesting and charge separation. Thus, they are very useful in artificial photosynthesis and generating hydrogen (clean source of energy) efficiently. It has been found that dendrimers are highly efficient with respect to other *n* number of layers [23].

Nanomaterials have been used in the synthesis of nanofuel cells made of a number of unit cells with each having a diameter of 200 nm. Each cell is formed by electrodepositing Pt-Cu nanowires inside cylindrical pores made of an anodisc filter membrane. Subsequently, Cu is dealloyed by soaking filter membrane into fumaric acid for several hours and finally constructing an array of porous platinum electrodes, with each having a diameter of 200 nm. It has been found that these nanofuel cells have power densities of c.1 $mW\,cm^{-2}$. These fuel cells can either be bundled in parallel or series without any complexities [24]. A fuel cell has been developed by using a nanocomposite [$LaNi_{0.2}Fe_{0.8-x}Cu_xO_3$ ($x=0.0$–0.2)] as cathode. The working range of this nanofuel cell lies in the temperature range of (500–700°C). This is one of the highest ranges of fuel cell operation at higher temperatures. Further, on doping with copper, the performance of the fuel cell has been enhanced by several hundredfold [25].

A novel ethanol fuel cell has been fabricated having an anode prepared from PtNiRu nanoparticles (1:1:1 atomic ratios) and a nanoporous TiO_2 film using a sol-gel and electrodeposition method. This ethanol fuel cell works by utilizing direct sunlight via a photocatalytic reaction on the anode. The presence of ethanol helps in electrooxidation under solar light and generates holes in the modified nanoporous TiO_2 film, which is responsible for excellent efficiency of the fuel cells [26].

Highly efficient nanomaterials for storage of hydrogen have been continuously evolving toward better efficiency. Different kinds of carbon arc discharge nanotubes have been checked for better performance in hydrogen storage. Samples taken include raw collaret collected on the cathode (1), raw soots collected on the lateral reactor wall (2), thermally treated soot (3), and thermally chemically treated soot (4). It was found that the last sample had maximum hydrogen storage at room temperature (approximately 0.13 H_2 wt%) [27].

A model based on grand canonical Monte Carlo simulation has been generated to elucidate hydrogen-storing capacity in nanoporous (primitive, gyroid, diamond, and quasiperiodic icosahedral) carbon materials and in CNTs. It has been found that carbon surfaces with varied geometry have limited hydrogen-storing capacity. However, additional additives like metals can help in efficient hydrogen storage [28].

Hydrogen gas is usually obtained by reforming hydrocarbons. However, CO is also produced in addition to hydrogen, which is toxic as well as limits the efficiency of fuel cell devices. A nanoparticle catalyst comprising a Ru core with its outer shell comprising a thick shell (1–2-monolayer) made of Pt atoms has been found to effectively remove CO from hydrogen through selective oxidation. The efficiency of this nanomaterial catalyst is quite high with respect to traditional catalysts like PtRu nanoalloys, monometallic mixtures of nanomaterials, pure Pt particles, etc. [29].

Hydrogen can also be produced by splitting water into its components. A number of devices are being developed for water splitting but none of them is completely effective. An efficient and stable nanostructured cell has been developed with its oxygen evolving anodes made of polyoxometalate cluster (totally inorganic ruthenium catalyst) with a conducting bed made of multiwalled carbon nanotubes (MWCNTs). The device has been found to be efficient and cost-effective, with a hope of producing large quantities of hydrogen and helping to make our planet free from carbon-based fuels [30].

Solar cells have been attractive candidates for producing clean and renewable sources of energy. Furthermore, photovoltaic cells based on nanomaterials are more efficiently and easily miniaturized, with easier incorporation into various electronic systems. Various innovations have appeared in this respect, as discussed here. Photochemical solar cells have been produced that contain an assembly consisting of single-walled carbon nanotubes (SWCNTs) and nanostructured SnO_2 electrodes having protonated porphyrins at their surface. This assembly is helpful in adsorbing sunlight in the visible range followed by electron ejection from excited porphyrins, as confirmed by femtosecond pump probe spectroscopy. It has been found that incident photon to photocurrent efficiency is approximately 13% with an applied voltage of 0.2V [31]. Photocurrent efficiency can be further improved by using three junction (InGaP/InGaAs/Ge) solar cells having Au nanoclusters deposited on their surface. Here photoconversion efficiency has been found to be 22.6% (much higher than any solar cells reported earlier) [32]. Two-dimensional nanostructures have been made using graphene, which have excellent electrical and material properties. This has brought faster electron transport with lower recombination due to higher light scattering [33]. It has been found that the higher the surface area for sunlight capturing, the higher is the photocurrent efficiency. Several innovations have appeared to enhance the surface of nanostructures being used in photovoltaic cells. Sponge-like structures have been made using a network of aggregated TiO_2 having very large surface area for interfacial reactions, such as charge carrier transfer routes. The morphology of this fabricated structure has been well characterized using atomic force microscopy, scanning electron microscopy, and field emission scanning electron microscopy. These techniques have confirmed advanced scattering and enhanced photocatalytic efficiency of sponge-like TiO_2–based nanostructures [34].

Hybrid nanostructures have been developed with improved photocatalytic capabilities. TiO_2 having a nanotube morphology has been doped with CdSe on its surface by using bifunctional linker molecules. This architecture helps in efficient electron injection from CdSe to TiO_2 upon bandgap excitation, which enables efficient photocurrent generation in a photoelectrochemical solar cell [35]. Nanostructures are molded into nanocrystalline sheets in the case of ZnO for synthesizing photovoltaic cells. It has been found that these structures exhibited better performance than other nanomaterial-based solar cells or well-oriented nanowire-based solar cells. These microstructures have increased incident light capturing with fewer losses during photoelectron transport, leading to better performance [36]. In recent years, even more innovations with respect to various forms of ZnO nanostructures have appeared to enhance the efficiency of solar cells. Comb-like nanostructures of ZnO were developed with fluorine-doped TiO_2 on their surface. The additional doping with structural modifications has led to enhanced light-to-electricity conversion by 0.68% with a fill factor of 34%, short-circuit current of $3.14\,mA\,cm^{-2}$, and open circuit voltage of 0.671V [37]. Various synthetic cationic dendrons were synthesized and doped on the surface of ZnO nanostructures, the idea being to come up with the most efficient dendrons with increased photocatalytic efficiency. In the current report, cationic dendrons such as 4-tert-butyl-1-(3-(3,4-dihydroxybenzamido) benzyl) pyridinium bromide (a), 1,1′-(5-(3,4-dihydroxybenzamido)-1,3-phenylene) bis (methylene) bis(4-tert-butylpyridinium) bromide (b), N1,N7-bis(3-(4-tert-butyl-pyridium-methyl) phenyl)-4-(3-(3-(4-tert-butyl-pyridinium-methyl)phenyl-amino)-3-oxopropyl)-4-(3,4-dihydroxybenzamido) heptanediamide tribromide (c), and N1,N7-bis(3,5-bis(4-tert-butyl-pyridium-methyl)phenyl)-4-(3-(3,5-bis(4-tert-butyl-pyridinium-methyl) phenylamino)-3-oxopropyl)-4-(3,4-dihydroxybenzamido) heptanediamide hexabromide (d) have been used for functionalization of ZnO nanostructures. It was found that the procedure thus adopted for loadings has resulted in efficiencies of 99.5%, 98.6%, 99.1%, and 42.5% with respect to (a), (b), (c), and (d) on the surface of ZnO nanostructures, respectively. Further, it has been found that option (a) was well suited for stable electrostatic attachment, leading to better photocatalytic efficiency with respect to the others [38].

Semiconductors have been the subject of experimentation for their efficiency in solar cells. The feature that makes them desirable in solar cells is their easier miniaturization and integration into power sources of nanoelectronic systems. It has been found that their higher efficiency is due to a higher amount of charge collection, carrier multiplication, and low-temperature processing with respect to other hybrid nanoarchitectures with poor stabilities. The p-type/intrinsic/n-type (p-i-n) coaxial silicon nanowires have been used for synthesizing solar cells and have yielded 200 pW per nanowire device using one solar equivalent illumination with a conversion efficiency of 3.4%. Further, these coaxial silicon nanowires can provide a new nanoscale test bed to study photoinduced energy/charge transport as well as being helpful in artificial photosynthesis. They are very useful in powering ultralow-power electronics and diverse nanosystems [39]. Hybridized Si solar cells have been made by using (CdSe) ZnS as core shell quantum dots of multicrystalline Si (2.4 nm) and tested for solar conversion efficiency. It has been reported that there is enhancement of conversion efficiency by two-fold under white light illumination [40]. Another hybrid silicon nanocrystal has been prepared by blending poly-3(hexylthiophene) (P3HT) polymer for synthesizing solar cells. In this case, direct solar illumination (100 mW cm^{-2}) has led to power conversion efficiency of 1.15% by using 35 wt% blended silicon nanocrystals [41].

Photovoltaics have been intensely researched to replace fossil fuels and research is still ongoing to enhance photovoltaic efficiencies. The major target is to use materials that can increase bandgaps and highly favor charge separation. It has been suggested that semiconductors are the best candidate in this respect, because their efficiency can be increased by blending with other semiconductors. Experimentation has taken place on blending silicon with germanium in two different ways: germanium has been blended onto the surface of silicon, while in the other case both silicon and germanium have been mixed and molded into nanowires. It has been found that the former combination has better performance than the latter; however, the photocatalytic efficiencies are higher than other reported nanostructure-based photovoltaic solar cells [42].

1.2.1.3 2011–2015

Solar cells based on CNTs have been developed for photoconversion. The best aspect of these solar cells is that they do not require rare source materials (In, Pt, etc.) or high-grade semiconductors. Further, no dye has been used for photoconversion, so there is no bleaching. Various optimization strategies have been carried out for improved efficiency of solar cells [43]. The transparent film of SWCNTs has been blended onto a conducting electrode made of Si:H of solar cells. This combination has led to improved performance due to removal of the Schottky barrier at SWCNTs/Si:H interface [44]. Various reports published on carbon-based solar cells have found enhanced photoconversion efficiency with respect to other metals and semiconductors.

Carbon is an excellent candidate for photovoltaic cells, because it has higher optical absorption, superior thermal and photostability characteristics, as well as the ability of reverting photodegradation. Further, solar cells based on carbon are cheaper, because carbon is one of the most abundant elements on earth. A photovoltaic cell has been developed that is composed of $PC_{70}BM$ fullerene, semiconducting SWCNTs, and reduced graphene oxide. This device has achieved a power conversion efficiency of 1.3% with very high photostability with respect to other carbon-based solar cell reports [45]. Hybrid carbon and inorganic nanocrystals have been used for synthesizing quantum dot sensitized solar cells. The electrode of these solar cells consists of carbon fiber (CF), Pt, and Co_9S_8 nanotube arrays (NTs). An absolute energy conversion efficiency of 3.79% has been demonstrated under $100\,mW\,cm^{-2}$ at 1.5 solar illumination. The device is very flexible with respect to substrates and is reported to have excellent energy harvesting and storage [46]. Carbon is best suited as a cathode, as suggested by various published reports. Research innovation is ongoing to obtain the best-suited anode for solar cells with excellent photoconversion ability. Hollow microspheres of SnO_2 have been shaped like cauliflower and tested for their performance in solar cells. Cauliflower-shaped hollow spheres of SnO_2 contain agglomerated nanomaterials that help in improved internal surface area and light scattering between shell layers. The structure is prepared by using successive ionic layer adsorption and reaction (SILAR) and chemical bath deposition (CBD) methods. It has been found that cauliflower-like SnO_2 structures are compatible for both quantum dot sensitized solar cells (QDSCs) and dye sensitized solar cells (DSCs). It has been reported that under 1 solar illumination, cauliflower structure-based solar cells have power conversion efficiency of 2.5% in QDSCs and 3.0% for DSCs [47].

Several biological species with vibrant colors either on their wings, shells, bones, or honeycombs have been studied thoroughly to understand the phenomenon of iridescent colors. It has been stipulated that such an understanding could help in various applications in addition to solar cells, such as in advanced sensors and photonic crystals. Complex natural architectures are responsible for regular, multiscale photonic structures present on a surface such as a butterfly wing, leading to production of dazzling vibrant colors. Various nanostructures are being fabricated that resemble these structures for different applications, including solar cells [48].

Another interesting innovative work on solar cells has found that, by using organic substances to make solar cells, biocompatible interfaces can be produced, such as for wearable electronics and human skin. Further, biocompatible organic solar cells are lightweight, flexible, and cheaper with respect to other solar cells. The silk fibroin has been used and integrated with a mesh of silver nanowires (AgNWs) (used in synthesis of solar cells). The power conversion efficiency of the solar cell has been found to be 6.62% with conductivity of $11.0\,\Omega\,sq^{-1}$ and transmittance of ~80% (in the visible light range). Furthermore, the cell retained conductivity even after being bent and unbent 200 times (a novel property not reported yet) [49].

Nanomaterials have been found to be useful in the synthesis of biodiesel. The basic mechanism of biodiesel synthesis involves transesterification of vegetable oil with alcohol, mostly methanol. Soybean oil has been most frequently used for biodiesel synthesis. Nanomaterials like ZrO_2 (10–40 nm) along with $C_4H_4O_6HK$ have been used as nanocatalysts for transesterification of soybean oil. The nanocatalyst has a long shelf-life, with sustained activity even after five cycles. Optimization has been carried out, to obtain a larger amount of biodiesel synthesis, with respect to molar ratio of methanol and oil, reaction temperature, amount of nanocatalyst, and reaction time. It has been found that a yield of 98.03% can be achieved using 16:1 (methanol to oil ratio) with 6.0% of nanocatalyst at 60°C kept for 2 h [50]. The heterogenous nanocatalyst $Cs_xH_{3-x}PW_{12}O_{40}/Fe\text{-}SiO_2$ has been prepared, using sol-gel and an impregnation procedure, and used for biodiesel production. A kinetic study of the transesterification reaction using Gc-Mas was done to calculate activation energy along with thermodynamic constants, including ΔG, ΔS, and ΔH, to elucidate the efficiency of the prepared heterogenous nanocatalyst with respect to other known catalysts. The heterogenous $Cs_xH_{3-x}PW_{12}O_{40}/Fe\text{-}SiO_2$ magnetic nanocatalyst can be recycled about five times with maintenance of catalytic activity [51]. Nanomaterials of activated Mg-Al hydrotalcite (HT-Ca) with a size of <45 nm have been prepared by an aqueous $Ca(OH)_2$ solution using coprecipitation and hydrothermal activation methods. HT-Ca has a high acid value (AV), so it can be used in both acidic and basic conditions. A biodiesel yield from *Jatropha* using 5 wt% HT-Ca has been found to be 93.4% at 160°C with methanol:oil of 30:1 having a reaction time of 4 h at AV of 6.3 mg KOH g^{-1}. Furthermore, a yield of 92.93% has been achieved in the case of soybean oil under similar conditions at AV of 12.1 [52]. Besides being used as a nanocatalyst in biodiesel production, nanomaterials can be directly mixed with normal diesel to improve the performance and emissions of diesel engines. The nanomaterials of cerium oxide (CeO) have been found to be effective in enhancement of engine efficiency due to enhanced oxygen buffering capability. In addition, it has been found that there is simultaneous reduction and oxidation of nitrogen dioxide and hydrocarbon emissions from the diesel engine [53]. Hydrogen is known as a clean source of energy and there is a focus on generating more hydrogen by various methods for its usage as fuel. CNTs could be very effective in generating hydrogen using waste glycerol from biodiesel production. It has been found that ~2.8 kg of CNTs is required to produce 500 Nm^3 H_2 using 1 ton of glycerol [54].

Fuel cells are promising electrochemical devices used to convert the chemical energy of fuels directly into electrical energy. Another research focus is on increasing the efficiency of fuel cells without generating any pollutants. Nanotechnology has helped in this respect, since a number of nanoparticles are known to increase the performance of fuel cells. Nanotextured copper has been used as an anode for fuel cells and tested for performance. It was found that nanotextured copper as an anode enhances the shelf life of fuel cells and leads to direct electro-oxidation of ammonia borane and additional amine derivates (efficient in converting chemical energy into electrical energy). Furthermore, this set-up reduced the total cost of the fuel cells [55]. Generally, fuel cells use a membrane for proton exchange. Slow reactions at the cathode are a frequent problem that affects the efficiency of the fuel cell. Several methods have been adopted to modify the cathode by using nanoparticles, the most common of which involves alloying platinum with 3d-transition metals like Fe, Co, etc. A hybrid structure has been synthesized by uniform doping of nitrogen onto MWCNTs blended with graphene oxide (N-G-MWNTs). The N-G-MWNTs hybrid nanostructure is used to enhance the catalytic efficiency of alloyed platinum (with transition metals) as a cathode of fuel cells. It was found that the performance of the fuel cell was five times higher than those without a hybrid nanostructure [56]. Another innovative approach for fuel cells is a membraneless design with flow-through electrodes that work by using several fuels like methanol, ethanol, glycerol, ethylene glycol in alkaline media, either individually or in mixed states. Furthermore, these fuel cells have a modified anode made of copper-palladium nanomaterials in the core shell, leading to outstanding performance along with other reported fuel cells [57].

Several innovations are in process to produce hydrogen economically and store it efficiently until its commercialization. However, hydrogen storage is not as easy as it seems, due to various challenging issues. The nanotubes of PdAg have been synthesized and tested for their ability of storing hydrogen. Results have shown that these nanotubes have excellent storing capability with almost no loss over several months. Further, pure Pd nanoparticles are not as efficient as PdAg nanotubes, with the latter having 200 times more storage efficiency than the former [58]. Another report has found that graphene, with a honeycomb lattice structure, is very efficient in storage/release of hydrogen under ambient conditions [59]. Hydrogen has been generated in the past using different chemical methods; the concept has now moved to producing hydrogen by artificial photosynthesis. Nitrogen-doped graphene oxide quantum dots have both p- and n-type conductivities. It has been found that they can be used for catalyzing water splitting (water is split during natural photosynthesis, generating electrons used for fixing CO_2) on visible-light irradiation. The quantum dots contain p-n type photochemical diodes with carbon (sp_2) clusters at their interfacial junction. The evolution of H_2 and O_2 takes place at the p- and n- domains of the quantum dots, respectively [60]. In another experiment, a hierarchical nanostructure of TiO_2 (3D hierarchical TiO_2, 1D/3D hybrid hierarchical TiO_2 composite, and 3D hierarchical protonated titanate microspheres) has been tested for its photocatalytic activities. Here the mechanism of formation of the hierarchical nanostructure of TiO_2, the reaction parameters, the morphology, and the crystal structure have been thoroughly studied and graded based on their water-splitting capacities. Photocatalytic activities have been determined by measuring generated hydrogen from water under UV irradiation in the presence of a sacrificial reagent (methanol). Self-assembled 3D hierarchical protonated titanate microspheres have exhibited much higher photocatalytic activity with respect to others due to increased surface area and enhanced charge carriers [61]. The CNTs have been hybridized with gold and used as an electrocatalyst for hydrogen production in alkaline media [62]. Subsequently, various other hybridized novel metals have been tested for improved efficacy in undergoing artificial photosynthesis. Most significantly, ultrathin CdS nanosheets as a photosensitizer and a nickel-based complex have been used as molecular catalysts for artificial photosynthesis. It has been found that there is an effective electron transfer from excited CdS nanosheets to the nickel-based complex, leading to efficient photocatalytic performance. The evolution of hydrogen was found to have a turnover of 28,000 and the lifetime of hybrid ultrathin CdS nanosheets with a nickel complex is over 90 h under visible light [63].

Various other substitutes have been explored for generating clean and green energy in addition to fuel cells and hydrogen generation. Lithium-ion batteries (LIBs) are best suited for energy storage due to having the least self-discharge rate. It has been found that the efficiency of LIBs depends on the electrode material. These batteries generally have lithium as their anode, because lithium can provide ions that can move rapidly between anode and cathode. LIBs are widely used in portable consumer electronic devices and electric vehicles ranging from full-sized vehicles to radio-controlled toys. Nanotechnology has been also useful in the synthesis of LIBs by providing a range of nanoparticles with peculiar properties that can be explored in the making of either cathode or anode. Nanostructured carbons like CNTs, carbon nanoparticles, graphene, and nanoporous carbon have been used as the anode of LIBs. It has been found that a nanocarbon network serves as an effective matrix for migration of electrons due to the structural stability and flexibility, leading to optimal battery capacity cycling stability and rate capability. It has been elucidated that nanocarbon networks are used as conductors and structural buffers, while noncarbon components like lithium are used for storage. Nanocarbon networks of CNTs and graphene are excellent candidates for use in LIBs. Furthermore, alloyed nanocarbon networks with Si and Ge have alleviated volume change, accelerated electron transport, and prevented agglomeration of nanoparticles [64]. The metal oxides like MnO_2, FeO_x, and RuO_2 have been tested for their performance in LIBs. LIBs based on these metal oxides have faster and reversible faradaic reactions, which are helpful in bridging the power/energy performance gap and thus enhancing charge-storage capacity while maintaining a few-seconds timescale of the charge/discharge response. Also, it has been found that when these metal oxides are electrodeposited onto carbon nanostructures, they promote effective electrolyte infiltration and ion transport to the nanoscale metal oxide domains within the electrode architecture, which further enhances high-rate operation [65]. Nanostructured sulfides (iron sulfides, nickel sulfides, molybdenum sulfides, tin sulfides, copper sulfides, cobalt sulfides, manganese sulfides, with zero-, one-, two-, and three-dimensional morphologies) are other potential candidates for use in LIBs, due to having excellent specific capacities. However, various practical complications can arise while using them in LIBs. Foremost are their poor cycling stabilities and inferior rate performance [66]. The most interesting innovation in LIBs is flexible mats made of SnxSb-graphene-carbon porous multichannel nanofibers prepared by an electrospinning method and subsequent annealing treatment at 700°C. These flexible mats have superior specific capacity of 729 $mA\,h\,g^{-1}$ in the 500th cycle at a current density of 0.1 $A\,g^{-1}$ with a reversible capacity of 381 $mA\,h\,g^{-1}$ at 2 $A\,g^{-1}$, which is much higher than those of nanofibers, graphene-carbon nanofibers, and SnxSb-carbon nanofibers. Further, these flexible mats are also useful in fuel cells and supercapacitors, in addition to LIBs [67].

Nanoparticles have been found to be very useful in generating energy using renewable resources like water and wind. A great deal of research has been done on making blades for wind or water turbines more efficient. The design of these blades significantly contributes to their performance. It is generally recommended to use very lightweight, thin, and elongated blades. Carbon nanofibers have been incorporated in fiberglass composites being used in wind turbine blades with a length of 40 m. Electricity generation was found to increase 100% when using 5% carbon nanofibers. Furthermore, their use has led to an increase in the lifespan of the blades. Several other combinations with varied percent of carbon nanofibers have been used and studied for blade performance using Monte Carlo analysis [68]. A focus has also been placed on the rotary radius of wind blades. The combination of CNTs with polymer nanocomposites has been ideal for increased efficacy of wind blades. Furthermore, this has led to cost reductions of several thousandfold. Studies have been done on wind blades made of CNTs/polymer nanocomposites with respect to mechanical, electrical, thermal, fatigue, and barrier properties. It was found that they are best suited for commercialization due to their excellent properties and cost efficacy [69]. In another interesting report, intensive research was carried out on making lightweight tower structures by integrating various types of nanostructured materials. The focus was on reducing weight by optimizing wall thickness of the tapered structure at the critical zones [70]. CNTs have been found to be very useful in synthesis of marine current turbines, which is one of the promising sources of clean and sustainable energy. Several studies have been done in this area, and the field has grown very fast in both industry and academia. However, very few reports have focused on utilizing nanoparticles in the synthesis of marine current turbines. The incorporation of CNTs into marine current turbines has produced very promising results. The parameters like structural reinforcement, fouling release coating, structural health monitoring, high performance wires/cables and lubrication have been well addressed [71].

1.2.1.4 2016–present

Plasmonic metal nanoparticles, like aluminum, are known to support localized surface plasmin resonance, which helps in boosting light absorption in solar cells. Furthermore, aluminum nanoparticles are cost effective, abundant, and are fully compatible with the metal oxide semiconductor (MOS) manufacturing process. Factors like shape of the nanoparticles, size, surface coverage, and length of spacing layer have been optimized to obtain excellent performance [72]. In previous decades, research innovations have focused on maximizing sunlight trapping for improved performance of solar cells. In this respect, a three-dimensional external light trap made of smoothened silver-coated thermoplastic facing towards the sun has been constructed, which retro-reflects the light via its structural parabolic concentrator and reduces optical losses while maximizing absorption over the large spectrum. It has been found that, by placing this trap on a thin nanocrystalline silicon solar cell, the efficiency of the solar cell increased by 15% [73]. A light trap has been prepared using gold nanoflowers around TiO_2 film present on the anode of solar cells. Gold nanoflowers have been found to have excellent light-harvesting efficiency, improved electron collection, and lifetime as well as slower charge recombination. They are especially useful for dye sensitized solar cells. It has been reported that short circuit photocurrent and open circuit photovoltage have been significantly enhanced, with a power conversion efficiency of 34% [74].

The lower olefins like ethylene, propylene, butylenes, etc. have been important raw materials for chemical industries. They are generally produced from range of hydrocarbon feedstocks like naphtha, gas, oil, light alkanes, etc. However, there are very limited stocks of hydrocarbons, especially petroleum, due to which alternatives must be found. The most common method is "Fischer-Tropsch to olefins" (FTO), which produces olefin by using syngas (mixture of H_2 and CO) derived from coal, biomass, and natural gas. However, the process, besides producing lower olefins (about 56.7%), also produces undesirable methane (29.2%). Thus, the process has not been very acceptable and is tedious. Nanotechnology has helped in this respect by using cobalt carbide quadrangular nanoprisms as catalysts. These catalysts have led to production of lower olefins (60.8%) with very little methane (5.0%) in FTO under mild conditions [75].

CNTs have been very useful in fuel cells as suggested from various reports. The porous CNTs, having a surface area of $502.9\,m^2\,g^{-1}$ with their surface containing MoO_2, rich in oxygen-containing functional groups, are used in sodium-ion batteries as anodes. It has been found that this has led to excellent rate performance and cycling stability ($110\,mA\,h\,g^{-1}$ after 1200 cycles at $5\,A\,g^{-1}$). Further, these porous CNTs are excellent candidates for other types of secondary batteries as well as being used as catalysts in fuel cells [76]. Three-dimensional structures have been prepared for PdCu nano-alloy and used as catalysts in hydrogen evolution and ethanol oxidation at alkaline pH. Amounts of hydrogen produced in this manner are much greater than previous reports [77]. Nanoparticles have been very useful in undergoing artificial photosynthesis and producing H_2 as well as O_2. Oxygen evolving complex (OEC), which is being used in natural photosynthesis, has been artificially developed containing CoO(OH) nanoparticles as the major component with CO_3^{2-} as cofactor. These CoO(OH) nanoparticles are the catalytic center, which helps by acting as proton acceptors from the O–O bond in the presence of CO_3^{2-}. The proposed OEC has been very useful in efficient electrocatalytic oxidation of water [78].

1.2.2 Environmental remediation

1.2.2.1 2001–2005

We all need clean air to breath, whether outdoors or indoors. It has been found that air is usually polluted in urban and industrial areas. Nanomaterials have been applied to treat polluted air in various ways. CNTs and fullerene, in multilayer shell-like structures, are able to collect emissions from natural gas flame and propane by undergoing their aggregation. Following aggregation, the polluted particles undergo thermal precipitation, leading to distorted and tangled structures taking the form of concentric rings collected inside the shell of nanomaterials and subsequently removed [79].

Biological degradation of hydrophobic organic contaminants like polycyclic aromatic hydrocarbons (PAHs) present in soil is not possible, due to their insolubility and poor mobility. It has been suggested to sequester them using polymeric nanonetwork particles made from a poly(ethylene) glycol modified urethane acrylate (PMUA). PMUA nanoparticles have been useful in effective sorption of hydrophobic organic contaminants as well as in their mineralization. Most importantly, adsorbed organic contaminants by PMUA nanoparticles have enhanced in-situ biodegradation rate in remediation through natural attenuation of contaminants. It has been indicated that these PMUA particles can be packed into bioreactors so that they can be recycled and undergo mineralization of organic contaminants on a large scale [80].

Dyes like azo dyes, acid orange 7 (AO7), procion red MX-5B (MX-5B), and reactive black 5 (RB5) have been environmental nuisances due to their nondegradative nature. It has been shown that their degradation is possible through nitrogen-doped TiO_2 nanocrystals. TiO_2 has been known for its photosorption; here doped TiO_2 nanocrystals have high reactivity under visible light. These doped nanocrystals can mineralize over 90% of dyes (azo dyes, AO7, MX-5B, and RB5) within 1 h, as shown by complete decoloration. The mineralization has also been confirmed by a decrease in total organic carbon and evolution of inorganic sulfate in dye solutions. These nanocrystals are much more efficient than any other known methods of dye mineralization [81].

Halogen-based compounds are not easily degraded naturally. However, nanotechnology has found solutions for this. Various halo-compounds have been treated in an efficient manner, as shown in several studies. CNTs pretreated with acidic solution have been used as adsorbents for trihalomethanes (THMs) from water. Here the CNT pretreatment is important as it has led to an increase in surface area and enhanced hydrophilicity towards low molecular weight as well as relatively polar THMs. The best pH range for adsorbing THMs from water by CNTs is from 3.0 to 7.0 [82]. An efficient degradation of anthropogenic chloroorganic compounds (COCs) like tetrachloroethylene (PCE), trichloroethylene (TCE), and carbon tetrachloride (CT) present in soil and groundwater can be carried out using nanosized metalloporphyrinogens immobilized in sol-gel. The degradation of COCs by nanosized metalloporphyrinogens takes place under reducing anaerobic conditions. Here, reusability of the nanoparticles is 12 successive cycles (an additional benefit). The degradation of COCs by nanosized metalloporphyrinogens can be enhanced by adding cyanocobalamin (vitamin B_{12}), which reduced the average time of degradation from 144 h to less than 48 h [83]. Nanoscale FeO particles have been used for in situ remediation of trichloroethene (TCE). Reaction conditions have been optimally standardized, particularly the synthesis of FeO nanoparticles and the level of H_2 used during the reaction. It has been found that the rate of TCE degradation by FeO nanoparticles is $92 \pm 0.7\%$ following pseudo–first-order kinetics having a rate constant of $1.4 \times 10^{-2}\,Lh^{-1}\,m^{-2}$. Most significantly, there is complete dechlorination of TCE without generation of any by-products such as acetylene and ethane, as in other methods [84]. Nanoscale zerovalent copper has been immobilized onto cation resin and used for the removal of carbon tetrachloride (CCl_4) from contaminated water. Here, cation resin has been used to prevent agglomeration of nanoscale zerovalent copper particles during degradation, which can lead to loss of efficiency; this technique is also helpful in reusing nanoparticles for several cycles. The degradation of CCl_4 has followed pseudo–first-order kinetics with a rate constant of $2.1 \pm 0.1 \times 10^{-2} Lh^{-1} m^{-2}$, which is several times higher than other reported methods of CCl_4 degradation [85]. The dechlorination of TCE and PCBs using bimetallic nanoparticles has been reported. The efficiency of degradation of TCE and PCBs by bimetallic nanoparticles has been improved by prevention of nanoparticle agglomeration. Here it has been reported that using palladized iron (Fe-Pd) nanoparticles in the presence of water-soluble starch as a stabilizer helps in prevention of agglomeration, with greater dechlorination efficiency. The starched nanoparticles at an approximate concentration of $0.1\, gL^{-1}$ and $1\, gL^{-1}$ have been used for transforming TCE ($C_0 = 25\, mg L^{-1}$) and PCBs ($C_0 = 2.5\, mg L^{-1}$) with efficiencies of 98% and 80%, respectively. Additionally, there are no toxic by-products generated during degradation of the PCBs and TCE [86].

The removal of hexavalent chromium [Cr(VI)] from water effluent is one of the tedious tasks during water treatment. The nanoscale maghemite has been prepared and used as an adsorbent for Cr (VI) from wastewater. Here factors like pH, temperature, initial concentration and presence of competing ions have been studied thoroughly. It was found that pH 2.5 is best suited for adsorption at 37°C. Further competition from metal ions like Na^+, Ca^{2+}, Mg^{2+}, Cu^{2+}, Ni^{2+}, NO^{3-}, and Cl^- are ignorable due to selective adsorption by nanoscale maghemite. The adsorption data has followed the Freundlich isotherm well [87].

1.2.2.2 2006–2010

Water contamination with chemical pesticides is currently a common problem due to increased pesticide use in agriculture. They are nonbiodegradable as well as not easily precipitable; thus they are not easily removable from water. Solutions of gold and silver nanoparticles have been used to remove chlorpyrifos and malathion (commonly found pesticides in surface waters of developing countries). Here, it has been found that the solution state of Au and Ag nanoparticles helps in precipitating the mentioned pesticides. However, the process of pesticide precipitation via nanoparticles takes more time. This problem can be easily sorted by immobilizing Au and Ag nanoparticles onto an alumina bed and passing contaminated water with pesticides through the bed with nanoparticles on its surface [88].

The presence of natural organic matter in drinking water treatment plants leads to formation of potentially harmful by-products, bacterial regrowth, and corrosion of pipes in distribution systems. A photocatalytic reactor has been made that contains stacked polymethylmethacrylate rings coated with TiO_2 nanofilm. The thickness of the rings has been optimally standardized to maximize the mineralizing efficiency of TiO_2 nanofilm. This TiO_2 nanofilm is able to mineralize natural organic matter as well as chlorinated by-product precursors like humic acids via photocatalysis through UV absorbance at 254 nm. The time needed for treatment of water containing humic acids and natural organic matter depends on their loadings. Further, the efficiency of the set-up for water treatment can be enhanced by using increased thickness of polymethylmethacrylate rings [89]. The photocatalytic capabilities of TiO_2 nanoparticles have been useful in degradation of various other water contaminants like methyl orange (MO) [90], 1, 4-dioxane (class 2B carcinogen) [91] and phenol [92].

TiO_2 nanoparticles have been useful in removal of heavy metals like pentavalent arsenate from water. This involves adsorption at an acidic pH that imparts a positive charge on the surface of TiO_2 nanoparticles. The adsorption isotherm follows the Langmuir model with maximum adsorption at pH 3.0, which is $8\,mg\,g^{-1}$. The presence of components like phosphate, sulfate, and bicarbonate are inhibitory to adsorption by nanoparticles [93]. Modified Fe_3O_4 nanoparticles with humic acid (11%, w/w, rich in O and N-based functional groups) with size of approximately 140nm have been found to be useful in removal of toxic heavy metals like toxic Hg (II), Pb (II), Cd (II), and Cu (II) from water. Most significantly, nanoparticles can be re-collected by applying low magnetic field gradients. The maximum adsorption for heavy metals of $79.6\,emu\,g^{-1}$ has been achieved within 15min with the sorption isotherm following the Langmuir model. The efficiency of modified nanoparticles in removing Hg (II) and Pb (II) is 99%, while that for Cu (II) and Cd (II) is over 95% in natural and tap water at optimized pH [94]. Nanoscale zero-valent iron has been used for removing almost all types of heavy metals present in wastewater or groundwater. Maximum remediation capacity has been found at pH 2.0 with concentration of nanoscale zero-valent as $2.0\,g\,L^{-1}$ [95].

Multiwalled carbon nanotubes (MWCNTs) are another potential candidate found to be useful in removing heavy metals like mercury (most toxic) from wastewater. Here, a slurry of MWCNTs has been prepared for removal of Hg^{2+} at a concentration of $1.0\,mg\,L^{-1}$ via an adsorption mechanism. It has been found that 10mg of MWCNTs is required to remove all of the $0.1\,mg\,L^{-1}$ of Hg^{2+} present in 50mL of wastewater at pH 4.0–8.0. Furthermore, increasing the mixing rate from 50 to 150rpm has led to tremendous enhancement in remediation efficiency [96].

The treatment of halo-compounds by nanoparticles has increased the significance of nanotechnology, as these compounds are some of the most nondegradable compounds known. Iron nanoparticles have been used for treating PCB (polychlorinated biphenyl) at 300°C in air, leading to a destruction efficiency of 95%. Furthermore, the addition of catalysts like V_2O_5/TiO_2 has significantly reduced the time of destructive ability of iron nanoparticles [97]. A nanofiltration membrane like NF200 and DS5 has been used to filter trihalomethanes and haloacetic acid (carcinogenic compounds formed from by-products of chlorine reaction with natural organic matter) from drinking water. The process is pressure driven, requiring utmost care for pressure application to avoid any rejection by the nanofilter membrane [98]. Nanoscale Cu/Fe bimetal particles have been used for dechlorination of hexachlorobenzene (HCB), which is a polychlorinated persistent organic pollutant. The report mentioned complete and fast degradation of HCB within 48h. It has been found that lowering of pH reduces the degradation of HCB by nanoscale Cu/Fe particles due to precipitation, which can be circumvented by adding iron oxide film on the surface of Cu [99].

Remediation of dyes from wastewater is another important application of nanoparticles. Semiconductor single-crystalline polar NiO {111} nanosheets having well-defined hexagonal holes are used as adsorbent for dyes like reactive brilliant red X-3B, congo red, and fuchsin red present in wastewater. The maximum adsorption as reported with respect to reactive brilliant red X-3B, congo red and fuchsin acid are 30.4 mg g^{-1}, 35.15 mg g^{-1} and 22 mg g^{-1}, respectively. Adsorption data fitted well with the Langmuir model and Freundlich isotherms; however, a better fit has been found with the Langmuir model [100]. In another report, reactive dyes like azo and anthraquinone have been removed from industrial wastewaters by using MgO nanoparticles via adsorption. Here, adsorption of reactive blue 19 and reactive red 198 are used as models for azo and anthraquinone dyes for optimizing dosages of MgO nanoparticles, dye concentrations, solution pHs and contact times. The maximum predicted adsorption capacities are found to be 166.7 and 123.5 mg of dye per gram for RB 19 and RR 198, respectively. The adsorption kinetic data followed a pseudo–second-order rate and fitted well onto the Langmuir model [101].

Nanocrystalline TiO_2, having a surface area of 329 $m^2\,g^{-1}$, surface site density of 11.0 sites nm^{-2}, total pore volume of 0.415 $cm^3\,g^{-1}$, and crystallite size of 6.0 nm, has been used to treat water containing uranium at pH $\geq$ 7.0. It has been found that the presence of inorganic carbonate has tremendously decreased the efficiency of nanocrystalline TiO_2 due to interference in complexation of uranium with TiO_2. The nanocrystalline TiO_2 is much more effective than any other known methods of uranium removal from wastewater [102]. Novel nanoparticles have been prepared by using magnetoferritin containing carboxyl groups at their surface derived from calixarene-crown-6 derivatives. They are used for decontamination of radioactive waste streams by sequestering radioactive cesium ions from water [103].

1.2.2.3 2011–2015

Negatively charged compounds are very difficult to degrade due to their resistance towards bioremediation by naturally occurring microbes. Various strategies have been adopted, but none of them has been completely effective. Nanoparticles have been useful in this respect, having excellent degradative capabilities for negatively charged compounds. The nanoparticles of FeO are homogeneously immobilized onto the surface of a nylon membrane and used for the reduction of nitrobenzene present in groundwater. Immobilized FeO nanoparticles are able to reduce the nitrobenzene content present in wastewater by 68.9% within 20 min at room temperature. Iron content present in the nanoparticles is mainly responsible for reduction of nitrobenzene. The reactions involving FeO nanoparticles and nitrobenzene have followed pseudo–first-order kinetics with their rate constants dependent on the pH of the solution [104]. Nanoscale zero-valent iron (NZVI) has been used for removing phosphate from wastewater. It has been found that NZVI has a phosphate removal efficiency of 87.01% in water containing phosphate (10 mg L^{-1}), when used at concentrations of 600 mg L^{-1}. Further, phosphate removal efficiencies of 99.41% and 95.09% can be achieved at pH values of 2 and 4 with concentration of NZVI as 400 mg L^{-1} in wastewater containing phosphate of concentration 20 mg L^{-1} [105].

The presence of nitrate in water has been a major detriment to the safety of drinking water. In this respect, FeO/Pd/Cu nanocomposites have been used for removal of nitrates from wastewater through catalytic and chemical reduction. Additional components like Pd and Cu have also been very useful, as they have helped in removing generated ammonia after nitrate reduction. The parameters like amount of FeO/Pd/Cu, initial nitrate concentration, solution pH, dissolved oxygen (DO), reaction temperature, the presence of anions, and organic pollutant have been properly optimized to obtain nitrate removal efficiency of 85% from water containing nitrate with concentration of $100\,mg\,L^{-1}$ at pH 7.1 [106].

Oxidized CNT sheets have been very useful in removal of heavy divalent ions like Cu^{2+}, Zn^{2+}, Pb^{2+}, Cd^{2+}, and Co^{2+} from wastewater via adsorption. The adsorption kinetics have found the preferential order of adsorption of divalent to be $Pb^{2+} > Cd^{2+} > Co^{2+} > Zn^{2+} > Cu^{2+}$ [107]. A hybrid nanomaterial (Poly 1, 8-diaminonaphthalene/multiwalled carbon nanotubes-COOH) has been found to be very effective in adsorbing heavy metals like Cd^{2+} and Pb^{2+} from wastewater. This hybrid nanomaterial has several extractive sites being used for adsorption. The maximum adsorption capacities for Cd^{2+} and Pb^{2+} are 101.2 and $175.2\,mg\,g^{-1}$, respectively. Furthermore, hybrid nanomaterial has been found to be useful in detecting Cd(II) and Pb(II) at concentrations of 0.09 and $0.7\,ng\,mL^{-1}$, respectively [108]. A novel reusable and cost-effective water purifier has been made containing nanocellulose (NC)-silver nanoparticles (AgNPs), which are embedded in pebbles-based composite material packed inside a glass column. The purifier is used for complete removal of dyes, heavy metals and microbes present in wastewater through adsorption (mainly through electrostatic interaction) and pore diffusion. The presence of porous concrete pebbles prevents leaching of AgNPs into treated water. This water purifier has shown 99.48% Pb(II) and 98.30% Cr(III) removal efficiency in addition to 99% decontamination of microbial load at an optimum pH of 6.0 [109]. A biomimetic sensor based on a self-assembled core satellite structure containing gold nanoparticles has been made to detect the level of Pb^{2+} in treated water. The sensor can detect 47.6 nM of Pb^{2+} present in treated water with excellent accuracy [110]. It has been found that nanomaterial based on titanate (*viz.* sodium titanate nanofibers and nanotubes) can be used for entrapment of radioactive cations, such as crystalline silicotitanate (CST), monosodium titanate (MST), and peroxotitanate (PT) present in wastewater [111].

Silver nanoparticles (AgNPs)-alginate composite beads have been synthesized and packed into columns. This set-up has been used for simultaneous filtration-disinfection of portable water. It has been very useful for water highly contaminated with microbes. This has been recommended for water treatment at point-of-use [112]. In another set-up, nanofibers have been derived from blends of grain proteins and polyethylene oxide for water filtration contaminated with microbes. These nanofiber-based filters are recommended for hospitals and senior residential areas [113]. Membrane fouling has been a common problem being faced with membrane filters for water treatment. MWCNTs/polyethersulfone (C/P) blend membranes have been prepared by using a phase inversion method. These C/P blend membranes are highly hydrophilic having higher pure water flux than polyethersulfone (PES) membranes without MWCNTs. The C/P blend membranes have a slower fouling rate, as shown in analyses of desorbed foulants [114].

Wastewater from petroleum refineries is highly contaminated and rich in organic pollutants. Remediation of this wastewater is very difficult; no technology is known to provide its complete remediation. Nanotechnology has been found to be very effective in this respect. Nanoscale-zero-valent iron (NZVI) has been used for degradation of effluents coming from petroleum refineries. Efficiency of degradation can be further improved by ultrasonication with optimized pH and dosage of NZVI [115]. Nanoscale tungsten oxide mesh has been used to prepare flower shaped structures having special wettability with superhydrophilicity in air and superoleophobicity under water. These flower shaped structures have excellent adsorption capacity and are useful for demulsification process. Being an emulsion breaker and separator, they are potential candidate in industrial fields such as water treatment and petroleum refining [116].

Scarcity of fresh water is one of the most alarming problems being raised on the stage of world. It has been hypothesized to desalinate sea water and use it for various domestic purposes and other applications. There are various technologies that have emerged in this respect with different claims. In addition to those, nanotechnology has also come up with effective water desalination. CNTs have been immobilized onto a hydrophobic membrane leading to enhanced membrane distillation, which helps in water desalination. Attachment of CNTs onto a hydrophobic membrane improves water-membrane interactions, thus promoting vapor permeability while preventing liquid penetration through the membrane pores. It has been found that this set up is able to desalinate water containing salt with concentration of 34,000 $mg\,L^{-1}$ at 80°C, effectively [117]. Nanoporous organosilica membranes coated onto porous alumina tubes are another successful example of water desalination via membrane distillation. Here, membranes produced pure water (up to 13 $kg\,m^{-2}h^{-1}$) by desalinating salted water with concentrations ranging from 10 to 150 $g\,L^{-1}$ at moderate temperatures ($\leq$60°C) [118].

Oil spills into water bodies have been very deleterious to aquatic ecosystems. The removal of oil from water is an extremely tedious task and the methods followed are not completely effective. Various reports have been published based on usage of nanoparticles for oil sorption, emulsification, and cleanup and revealed their efficacies with respect to other known technologies. Nanohybrids having an onion-like shape have been prepared containing SWCNTs or MWCNTs fused to silica or alumina particles. These nanohybrids are used for oil emulsification leading to small oil droplet formation having a size <40 μm. It has been recommended that the addition of transition metal particles like palladium and copper to nanohybrids helps in catalyzing reactions at the water/oil interface, *viz.* hydrogenation of phenanthrene, hydrogenation of glutaraldehyde and benzaldehyde, oxidation of tetralin, etc. [119]. Nanoporous polystyrene fibers have been found to have high oil sorption capacity and they are cost effective with excellent selectivity. The nanoporous polystyrene fibers have sorption capacities for different oils as: 113.87 $g\,g^{-1}$ (motor oil), 111.80 $g\,g^{-1}$ (bean oil) and 96.89 $g\,g^{-1}$ (sunflower seed oil), making them a potential material for oil sorption in wastewater treatment [6]. CNTs have been implanted into porous ceramic channels and used for removing tiny oil droplets from water. CNTs are hydrophobic in nature and possess an interfacial curvature at nanoscale that helps in trapping tiny oil emulsions, even with sizes at submicrometer, and allowing water to pass through. CNTs are uniformly and optimally distributed onto ceramic to obtain uniform lipophilic layers during filtration, for improved performance. It has been found that there is 100% oil rejection when the permeation flux is $0.6 \mathrm{L\,m^{-2}\,min^{-1}}$ with pressure of 1 bar for 3 days as shown by using wastewater containing 210 ppm of mineral oil, 1600 ppm of emulsifier, and a trace amount of dye [120]. Polyvinylidene fluoride (PVDF) membranes (MM) are fabricated with nanosized TiO_2/Al_2O_3 and used for separating oil from water. Here the pore size of the modified PVDF membranes has been narrowed to increase hydrophilicity and decrease contact angle. This set-up for oil separation is very fast, cost effective, and has easier adaptability than existing water purifiers [121]. Thermally reduced graphene has been used for oil-spill cleanup through adsorption. It has been found that it has a high sorption capacity of 131 g of oil per gram of thermally reduced graphene, which is higher than any other carbon-based sorbents [122]. In another report, thermally reduced graphene has been used to prepare graphene oxide foams using three freezing methods: unidirectional freezing drying (UDF), nondirectional freezing drying, and air freezing drying. It has been found that reduced graphene foams prepared by UDF have maximum adsorption capacity of $100 \mathrm{g\,g^{-1}}$ for all oils (gasoline, diesel oil, pump oil, lubricating oil, and olive oil) with the highest value corresponding to olive oil ($122 \mathrm{g\,g^{-1}}$) [123]. In another report based on graphene adsorption of oil and organic solvents, magnetic graphene was prepared by depositing Fe_3O_4 nanoparticles and used for adsorption. The reduced graphene foam film prepared by gaseous reduction in the hydrothermal system with deposition of Fe_3O_4 nanoparticles on their surface provides them porous morphology in addition to magnetism. The porous structures with magnetic properties have outstanding oil adsorption capacity, excellent recyclability, and stability under cyclic operations to graphene foam for oil spill cleanups [124]. Photoresponsive TiO_2 has been used for removing industrial oil waste and thus is helpful for remediation of oil spill pollution from water bodies. A nanosponge has been made composed of hydrophobic hydrocarbon and hydrophilic TiO_2 nanoparticles for oil absorption or desorption in response to UV irradiation. The hydrocarbon in the nanosponge helps in oil absorption from water followed by release of absorbed oil into the water by TiO_2 in response to UV irradiation. It has been found that functionalization of the nanosponge with polydimethylsiloxane further helps in release of more than 98% of the absorbed crude oil upon UV irradiation. This is a smart technology for oil absorption/desorption with excellent selectivity and recyclability, with the additional advantage of reuse of the spilled oil [125].

There are reports based on remediation of air pollution through nanoparticles. The results obtained thus far have exceeding those of various existing technologies. In one of the reports, an existing air filter has been modified by using an antimicrobial nanoparticle (produced by *Sophora flavescens*) coated onto an electrostatic air filter. This air filter has ~92.5% filtration efficiency (for 300 nm KCl aerosol) and

>99.0% of antimicrobial activity with minimal pressure drop (~0.8 mmAq at a flow rate of 13 cm s^{-1}). This air filter has been recommended for the indoor environment for various functions [126]. A novel electreted polyetherimide silica (PEI-SiO_2) has been blended onto a porous membrane for undergoing air filtration and purification. The filter prepared in this manner has superhydrophobicity in addition to excellent self-cleaning performance after filtering the air. The filter purifies air through its porous structure and electric charge, having a filtration efficiency of 99.992% with a pressure drop of 61 Pa in 30 min at 200°C. It has been anticipated that this filter can be used as a core part of various air filtration systems, like ultra-low penetration air filters, clean rooms, respirators, and protective clothing [127].

An interesting report has been published based on introducing metallic nanoparticles into diesel engines and reducing emission of toxic air pollutants. Here, silver nanoparticles have been found to be very effective when added into diesel fuels. The presence of silver nanoparticles in a diesel engine helps in altering engine power, oil temperature, and subsequently the proportion of released air pollutants. It has been found that the rates of CO and NOx have been reduced by 20.5% and 13%, respectively. Further, this technique has also modified hydrocarbons by up to 28% in addition to minimizing fuel consumption and improving engine power by severalfold [128].

1.2.2.4 2016–present

The addition of antibiotics into water has usually been followed by traditional water treatment. However, it is essential to remove those antibiotics before releasing water for usage, especially for drinking. A dynamic system based on graphene oxide (GO)/nanoscale zero-valent iron (nZVI) has been produced to adsorb and oxidize antibiotics present in water at specific pH. This system is able to work even in the presence of high salt and humic acid, with a working pH range from acid to neutral. Once water is treated, the GO/nZVI composite can be easily removed by applying a magnetic field and regenerated by keeping it in an alkaline solution (pH > 9.0). GO/nZVI is a promising candidate for water treatment due to its high removal efficiency, high stability, reusability and easier separation [129]. An electrospun carbon nanofiber carbon nanotube (CNF-CNT) composite has also been used for water treatment, especially useful for removing organic micropollutants. Its kinetics, isotherm and pH-edge sorption studies were carried out using sulfamethoxazole and atrazine. This system is recommended for point-of-use (POU) water treatment based on its sorption capacities and mechanical strength [130]. The potent toxins like cyanotoxins, microcystins, and cylindrospermopsin produced by cyanobacteria present in potable water supplies can be easily removed by magnetophoretic nanomaterial via adsorption. The adsorption process has been found to be pH dependent and follows pseudo–second-order kinetics having the Langmuir isotherm as an adsorption model [131]. A polymer-metal oxide nanofiber (PAN) functionalized with surface active quaternary ammonium salts has been used as a filter for POU water treatment containing metal oxyanions (*viz.* arsenate, chromate). Iron oxide (ferrihydrite (Fh)) nanomaterials embedded inside nanofiber can improve metal-adsorbing capacity. Kinetic studies of hybridized PAN have revealed that chromate and arsenate are removed through ion exchange; however, the former is more susceptible to interference by counter ions [132]. A nanoporous two-dimensional Fe phthalocyanine (FePc) membrane has been used for desalination of ocean water. The FePc membrane conducts a fast flow of water and suppresses ion permeation through exclusion of cations due to electrostatic repulsion and trapping anions via its narrow pore size. The number of protonated nitrogen atoms present in the FePc pores are modulated by pH adjustment, which in turn helps in regulating anion occupancy, giving rise to control of water flow. The FePc membrane was found to have excellent selectivity, permeability, and controllability [133]. Also, a nanocatalyst polyvinylidene fluoride/polyacrylonitrile (PVDF/PAN) composite grafted with acrylic acid and decorated with metal nanoparticles (Ag, TiO_2, Fe-Pd) has been used for treating water contaminated with toxic pesticides like dieldrin, chlorpyrifos, diuron, and fipronil through their dechlorination. Experimental results have found that dechlorination of pesticides followed first-order kinetics with efficiencies of 96%, 93%, 96%, and 90% with respect to dieldrin, chlorpyrifos, diuron, and fipronil at 5 ppm, respectively, after 3 h of treatment [134].

Fine particulate matter (PM) present in the atmosphere has affected human health and climate adversely. The smaller the size of the particulate matter, the greater are its adverse effects. The $PM_{2.5}$ (mass of particles below 2.5 μm in diameter) concentration present in the atmosphere can be reduced effectively through polyacrylonitrile (PAN) nanofibers modified with ionic liquid diethylammonium dihydrogen phosphate (DEAP). The modification has improved surface roughness of the nanofibers and thus increased the number of adsorption sites for capturing $PM_{2.5}$. [135]. A high-efficiency (>99.5%) polyimide-nanofiber air filter has been synthesized for removing $PM_{2.5}$ over a broad temperature range (25–370°C). These filters have high air flux with very low pressure drop and can work continuously for more than 120 h. These filters are recommended for car exhaust due to their high efficiency and working capacity at higher temperatures [7]. Nanofiber-based air filters made of chitosan have been used for capturing $PM_{2.5}$ with a high efficiency through electrostatic precipitation. It has been shown that this nanofilter has a removal rate of $PM_{2.5}$ of 3.7 $\mu g\,m^{-3}\,s^{-1}$ with an efficiency of >95% $PM_{2.5}$ as obtained from a field test in a smoking room (~5 m × 6 m × 3 m). Moreover, chitosan used for synthesizing nanofilters is nontoxic, biodegradable, and harmless to human health [136].

Electrospun carbon nanotube/titanium dioxide (CNT/TiO_2) nanofibers have been used to degrade gaseous benzene through photocatalysis under visible light irradiation. The CNT/TiO_2 nanofibers are fabricated with poly(vinyl pyrrolidone) (PVP) for improvement of specific area and photocatalysis of benzene at 100 ppm. The large surface area and lower bandgap energy of CNT/TiO_2 nanofibers helps in stronger visible light adsorption, thus leading to improved air treatment efficiency using less energy [137]. The PAN nanofibers containing MgO nanomaterials have been prepared with the porosity best suited for air filtration. This filter has efficiency higher than a high-efficiency particulate air (HEPA) filter with excellent collection efficiency and minimal pressure drop. This filter has been recommended for filtering industrially polluted air through adsorption, pore size, and antibacterial functionality [138]. Nanomaterials like TiO_2, CNT, and V_2O_5 have been recommended to coat carriers used for air filtration. These nanomaterials are found to be effective catalysts in removing toxic air pollutants like PCDD/Fs (polychlorinated dibenzo-*p*-dioxins and polychlorinated dibenzofurans) and NOx (nitrogen oxides) through adsorption. The removal efficiency of PCDD/Fs by V_2O_5/TiO_2-CNTs is 99.9% at 150°C [139].

1.3 Nanotechnology: a solution link for the energy crisis and environmental pollution

The advantages of nanomaterials are quite diverse; in fact they have possible applications in almost every aspect of our daily lives due to their versatile properties. Different US federal agencies invested over $1 billion at the beginning of the 21st century, just when the field began to emerge with many applications. Industrial investment is still growing, with a number of nanomaterials already having landed in the commercial market. For example, titanium dioxide (TiO_2) nanomaterials are marketed by Proctor and Gamble in sunscreen lotions to shield skin from UV radiation. Nanomaterials are being used in sporting equipment, clothing, telecommunication infrastructure, and fuel cells, among others. The field became even more popular when nanotechnology applications for solving various environmental and energy problems came into the limelight. Nanofilters are now available that can make even saline water fit for drinking cost-effectively, helping to solve water scarcity problem in arid regions. With the invention of principal solar cells and solar paints based on nanoengineered materials, dependency on fossil fuels can be lessened. Nanostructured aerogels are known to be extremely strong inexpensive materials that are helpful in making structures resistant to earthquakes and hurricanes. Nanomaterials are important components in sensors used for detecting extremely low concentrations of contaminants in water, air, or soil with utmost accuracy. Even heavy metals and microbial toxins can easily be detected in drinking water within microseconds using nanoscale sensors. Nanotechnology has not only been helpful in detection of contaminants but also in their quantization and remediation. Iron oxide nanoparticles encapsulated in a protein shell are being used for the reduction of heavy metals, such as hexavalent chromium (frequently found in groundwater). Even toxic recalcitrant water contaminants like trichloroethylene (TCE) can be reduced per milligram Fe per hour three orders of magnitude better than any other conventional technology being used [140]. There are many more examples, which are discussed in coming chapters. Nanotechnology undoubtedly has great potential in its ability to help clean up the environment with greater efficiency and cost efficacy.

Nanotechnology has been very useful in generating green and clean energy (efficiently producing energy while eliminating polluting waste products). The most common examples are solar and fuel cells being used in various appliances and automobiles. Research is still evolving in nanoscale photosynthesis, microfluidic biofuel cells, photoelectrochemical cells, and electrochemistry to address the growing need for safe, inexpensive, and renewable energy resources. LEDs (light-emitting diodes) based on nanocrystals/polymer composites are found to be brighter, more efficient, and less costly. Further, usage of LED technology can reduce power consumption by 50% and thus reduce carbon emissions by about 28 million metric tons per year, saving $100 billion in 10 years. The introduction of hydrogen cells based on nanomaterials have revolutionized the concept of clean energy, as the burning of hydrogen produces water, which is completely safe for the environment. It has been stated that motor vehicles using hydrogen fuel would not only save our planet from global warming but would also completely solve the problem of fossil fuel scarcity. Further, lighter-weight batteries based on nanomaterials have lowered the amount of fuel needed for transportation and decreased the usage of electricity in various electrical appliances. Most importantly, the usage of nanocatalysts has led to cleaner, more cost effective, and more eco-friendly petroleum refining, which has made nanotechnology more profitable.

Nanotechnology's development has been a boon due to its vast applications with additional economic benefits, particularly its advantages in environmental remediation and production of clean energy, as previously described. However, the unintended effects of nanomaterials that can adversely impact the environment, both within the human body and within the natural ecosystem, have been a serious issue [141]. It is an utmost necessity to identify the negative aspects of this technology before it is widely introduced into the marketplace. Very few studies have focused on the effects of nanomaterials on human health and the environment. Furthermore, several questions have been raised about the genuine applications of nanotechnology rather than speculations based on market statistics. The following points provide a brief overview of the negative prospects, which have been overlooked in the profusion of publicity and highly rated reports on applications of nanotechnology.

- *Manufacturing nanomaterials has huge energy and environmental costs.* Nanotechnology has provided a wide range of nanomaterials with a number of promising applications, including solutions to environmental problems. These materials are known to lower energy consumption in vehicles like cars, trucks, airplanes, boats, and even spacecraft; they show promise in reducing energy loss during transmission and can make energy production from renewable sources like the sun and wind more efficient. They can also be an important substitute in storage batteries based on toxic heavy metals and they are helpful in increasing the lifetime and efficiency of storage batteries. However, manufacturing nanomaterials has adverse environmental fallout for various reasons. It involves a large amount of energy and water consumption, a large amount of waste generation, production of greenhouse gases such as methane, and usage of toxic chemicals and solvents like benzene. Reports have found that their environmental and human toxicity is about 100 times higher per unit of weight than those of conventional materials like aluminum, steel, and polypropylene. Nanomaterials like nanotubes have been proven to be quite noxious to the lungs and may even lead to cancer (RS/RAE 2004; Swiss Re 2004). A series of experiments has also found that they can cause inflammation, granuloma, fibrosis, arterial plaque with subsequent heart attacks, DNA damage, etc. Furthermore, they are important contributors in ozone depletion due to generation of greenhouse gases during their large-scale production [142]. Thus, their environmental gains may be outweighed by environmental toxicity and expensive cost of production.

- *Many commercially used nanomaterials are toxic to the environment.* A correlation has been drawn that says that the lower the size of the particle, the higher will be the production of radicals, thus the higher would be the toxicity. Laboratory studies have found that nanomaterials have potential DNA-damaging activity that directly affects metabolic activities of living organisms and can even lead to death. The Woodrow Wilson International Center for Scholars' Project on Emerging Nanotechnologies (PEN) in 2006 predicted that there would be about 58,000 metric tons of nanomaterial production worldwide in the period 2011–2020. This has huge impact on environment which is several thousand folds higher than any other known conventional materials. Lab-based studies have confirmed the nanomaterial toxicity on the environment by using standard indicators like algae, invertebrates, and fish species. Some nanomaterials are known to impair reproductive cycles of earthworms (a major component in nutrient cycling), thus disrupting functioning of the ecosystem [143]. Carbon nanomaterials (C70 fullerenes and MWCNTs) have been very deleterious to important staple crops (rice and wheat) by delaying their flowering period by over 1 month, which subsequently lowers their productivity by severalfold. It has also been reported that they can pierce the cell wall of wheat plants' roots, making easy channels for transportation of pollutants into the living cells [144]. Further, they are important biomagnifiers due to their nondegradative nature.
- *Solar energy achievements remain incremental, while giving rise to concerns in usage of toxic nanomaterials.* Nanomaterials like titanium dioxide, silver, quantum dots, and cadmium telluride are known to be cost-effective alternatives for synthesis of solar panels, with efficiency about 60% higher than traditionally used wafer-based crystalline silicon solar panels. Data provided by Nanosolar (a US firm) on economic production of solar panels based on nanomaterials shows US$1/watt, while a European firm has found US$0.50/watt. Further, nanosolar panels are useful in reel-type label screen printing, energy-generating photovoltaic paint converting infrared radiation into electricity, etc. However, the toxicity level of the mentioned nanomaterials used for solar panels is highly potent, which raises more environmental issues [145].
- *Concerns have arisen over water decontamination technologies using nanomaterials.* Nanotechnology-based water treatment technologies using nanomembranes, nanomeshes, nanofibrous filters, nanoceramics, nanoclays, zeolites, and nanocatalysts have been found to be very efficient, cost effective, durable, and much faster than available conventional technologies. However, issues like safety concerns, sophisticated technological capabilities, and much cheaper local water treating technologies available (*viz.* a four-times folded old sari can remove 99% of cholera bacteria) have been major reasons

for noncommercialization of nanomaterial-based water treatment. Even benefits like allowance of salts and other necessary substances important for health have not lured the common masses. Companies like General Electric, Dow Chemicals, Siemens, etc. have claimed to earn billions of dollars from nanotechnology-based water treatment; however, the technology is still far away from the common people.

1.4 Health and safety issues in nanotechnology

With today's increasing health awareness programs and heightened health concerns, it has become increasingly important to release all the details of any major new product on the market. Nanomaterials have become a crucial part of many applications affecting our daily lives, including environmental remediation and production of clean energy. Health agencies must take responsibility for providing thorough and unbiased details on any nanomaterials before they come into the marketplace. Publications analyses on nanomaterials have often found that there is not even a single mention of toxicological studies. There appears to be a strong bias toward reporting only the positive aspects of nanomaterials. Therefore, a real need exists to start from scratch on toxicological research, with backup from government and other funding agencies. While such studies are being carried out, it is also important to mention accurate laboratory conditions, as entirely different effects can be found when nanomaterials are released into the environment, with much more complex conditions. Such studies should search for different methods to nullify the effects of nanomaterial toxicity. Obtaining a clear picture of the toxicity situation will be a boon, as opposed to hiding the details in order to bring a product to market. Some studies appeared a few years ago on neutralizing nanomaterials toxicity through surface modifications made to alter their biological interactions. It has been hypothesized that surface modifications can even make it possible for humans to eat dangerous nanoscale mercury. The surface coating has a significant effect on the environment and health, because it is the most important region directly interacting with various biological moieties. Before going into more detail on surface modifications, it is important to understand the toxicological effects of nanomaterials on biological systems through various mechanisms.

- *Interactive ability with biological systems*: The foremost issue on everyone's mind is the interactive capability of nanomaterials with biological systems. One cannot generalize the effects based on studies on just a few nanostructures, as there are other factors responsible for interactive capabilities besides the small size. The small size, of course, is a great aid in gaining easier access into any metabolic or cellular activity. The size of nanomaterials makes them highly reactive systems at their surface and can lead to generation of free oxygen radicals, which are responsible for many diseases and pathological conditions [146]. The type of nanomaterial is what actually decides the type of interactions and effects that occur. However, it has been generalized that a size smaller than 50 nm is highly toxic due to the easier translocation into cells, and the resultant greater ability to interact with channels, enzymes, and various other cellular proteins [147].

- *Administration routes and health effects*: There are very few studies on the health effects of the route of nanomaterial administration. During the early days, when nanomaterials had just come into existence, it was hypothesized that their mechanism of action on living organisms would follow a similar path to that of viruses. It has been found that nanomaterials can easily bypass the blood-brain barrier, which is perhaps the most important matter of concern [148]. In the case of fullerenes, it has been found that their toxicity is directly related to the dosing route. Oral administration of fullerenes is the least toxic, with elimination of over 98% of fullerenes through feces and urine within 48 hours, while 2% is still found in the body at that point. In the case of intravenous dosing, however, fullerenes are transported rapidly into various body parts (liver: 73%–92%, spleen: ~2%, lung: ~5%, kidney: ~3%, heart: ~1%, and brain: ~0.84%) within 3 h. Even after 1 week, 90% of intravenously administered fullerenes are still found throughout the body [149]. Due to the small size of the nanomaterials, they can be easily inhaled and enter the respiratory tract. They are found to easily evade alveolar macrophages and enter the lymphatic and circulatory systems within a very short time span. The problem becomes more severe when inhaled nanomaterials are transported to the brain via the olfactory or trigeminal nerves. However, not all nanomaterials are found to enter the respiratory tract despite their very small size. For example, only 20% of nano-TiO_2 can enter lung alveoli due to unknown reasons. Here, very high dosage rates are required to study the health toxicity of nanomaterials [150].
- *Oxidative stress*: This is a major issue needing to be seriously addressed, because oxidative stress is a major cause of cell damage and eventually cell death. Oxidative stress can be diagnosed by various methods: DNA damage, inflammatory response, brain translocation, level of glutathione (GSH), protein oxidation, lipid peroxidation in organs (brain, liver, gills), etc. The genes associated with hormone regulation, immune cell response, and clotting and anticlotting pathways are also known to be altered during oxidative stress brought on by nanomaterials. The levels of repair enzymes such as superoxide dismutase increase significantly to overcome the damage caused by nanomaterial-related oxidative stress. Nanomaterials like C60 fullerenes, metal Qdots, and TiO_2 are found to be highly redox active and to produce singlet oxygen and superoxide radicals within a very short time after their introduction inside the body [151]. Studies have been done on the chemical nature of nanomaterials and their correlation with generation of reactive oxygen species, which is directly related to oxidative stress. Generally, ultrafine organic carbon-based nanomaterials are more toxic than inorganic or metallic nanomaterials. They are responsible for a higher rate of glutathione depletion and mitochondrial damage in directly exposed cells. Oxidative stress leads to proinflammatory effects like allergy, asthma, and even propagation of cardiovascular disease. Experimental animals exposed to potential organic

nanomaterials have had enhanced allergic airway responses and CNS (central nervous system) inflammation with significant decreases in heart rate. Further, there is additional risk of preterm birth and low birth weight [152].

- *Inhalation and cytotoxicity*: Particle size/determines whether the particle will be respirable or inhalable. In the case of a rat model, a size smaller than 3 μm is respirable while above this size it is inhalable; for humans, a size of 5–10 μm is respirable and 10–50 μm is inhalable [153]. Thus, nanomaterials are potential agents that can easily be inhaled and deposited at the junctions of the bronchioles and alveolar ducts, leading to obstruction in gaseous exchange. The state of nanomaterials is significantly important; *viz.* aggregating nanoparticles are less toxic than discrete, fine-sized nanoparticles. Heterogenous nanoparticles are more prone to aggregation than fine-sized particles of identical composition. Surface properties are the major determinant for the aggregating capabilities of nanomaterials. Recent work on CNT toxicity has revealed various facts. CNTs have the inherent capacity of aggregating into nanoropes due to their strong electrostatic potential. Following the instillation of CNTs into the lung of rat, histopathological studies, cell proliferation, and lung weight measurements were done within 24 h, 1 week, 1 month, and 3 months postinstillation. It was found that 15% of the animals died within 12 h due to choking from the high rate of agglomeration from the strong electrostatic attractions. Here, the nanomaterial itself was not toxic but strong electrostatic interactions were the crucial factor. Multifocal granulomatous lesions were observed in the lung tissues of survived rats, with irregular lesion distributions. The center of the lesions contained agglomerated carbon nanoropes surrounded by foamy multinucleated macrophage cells, indicating that agglomeration takes place inside alveoli rather than in the respiratory tract, which prevented instant death of the rats. Here the rate of dosage has no role, as found with administration of different concentrations of CNTs. Therefore, case by case studies must be done with different nanomaterials, rather than generalizing the same effects for different situations and materials [154]. However, the surface properties play a crucial role in any nanomaterial-based disease pathogenesis. Studies should focus on the properties such as surface coating, surface charge, dispersion, and particle aggregation, which would help in assessment of the accurate toxicity of any given nanomaterials.

1.5 Nanotechnology and strategies to ensure occupational health

The debate on the toxicity of nanomaterials has been heating up and several points are being raised. The most important point is the safety of the people working on nanomaterials, because they are the first to encounter them; the problem becomes even more severe if someone is working closely with novel or engineered nanomaterials. Nanomaterials are heterogeneous in nature, insoluble, and have surface properties entirely different from the core properties. Due to their very small size, nanomaterials are able to gain easier entrance into the respiratory tract and subsequently other organs. Fig. 1.4 shows an overview of nanomaterial toxicity through their easy entrance inside the body. In the case of humans, extent of toxicity can still be done but there are no validated protocols for carrying toxicity effects on plants. Various strategies have been planned to nullify the toxic effects of nanomaterials on the environment, but none of them has yet been fruitful due to a variety of reasons. Most importantly, the focus has been only on the size of the nanomaterials and the toxicological effects related to size, when their chemical reactivity towards various biological metabolical activities has a more crucial role in toxicity [155]. There is no protocol for neutralizing the toxic effects of nanomaterials while using them in various applications. The major problem is inadequate information about the physical and chemical properties of natural or engineered nanomaterials. Some efforts are being made to study human toxicity, but there are few concerns at this point about environmental toxicity. Strenuous work needs to be done by various collaborations worldwide to come up with suitable solutions. There should be strategic planning with clear motives for short, medium, and long-term goals for coordinated activities. The problem should be addressed from various perspectives, including industry, the public, workers, and regulatory authorities.

- *Regulatory framework*: There should be a well-defined regulatory framework for the release of nanomaterials for various applications, as is done with the release of other new chemicals, pesticides, cosmetics, pharmaceuticals, etc., through various regulations like TSCA (Toxic Substances Control Act). The basic information should be obtained, such as physico-chemical properties, toxicity using standard cell lines and model organisms, environmental impact on release into the environment, etc. These require additional studies and experimentation, which could delay the commercialization of nanomaterials, but would be much safer. The United States Environmental Protection Agency (US EPA) has generated a number of norms related to release of toxicity data, which is the major impedance for commercialization of nanomaterials. In the case of the European Commission, their algorithm is much easier, which has rather helped in bringing nanomaterials to market. Their decision algorithm includes solubility, ability to be transported within the body, and probability of ecotoxicity and subsequently prioritizes these as low, intermediate, and high priority, to derive a research agenda. However, here the point is not for an easier pathway for nanomaterial commercialization but its toxicity to humans, other living organisms, ecosystems, and subsequently our environment. A most important question is: how will implementation and enforcement of these algorithms be handled by the manufacturers?
- *Moving forward*: A major question aimed at our scientific community is whether to move forward, or not. How should we proceed in order to take full

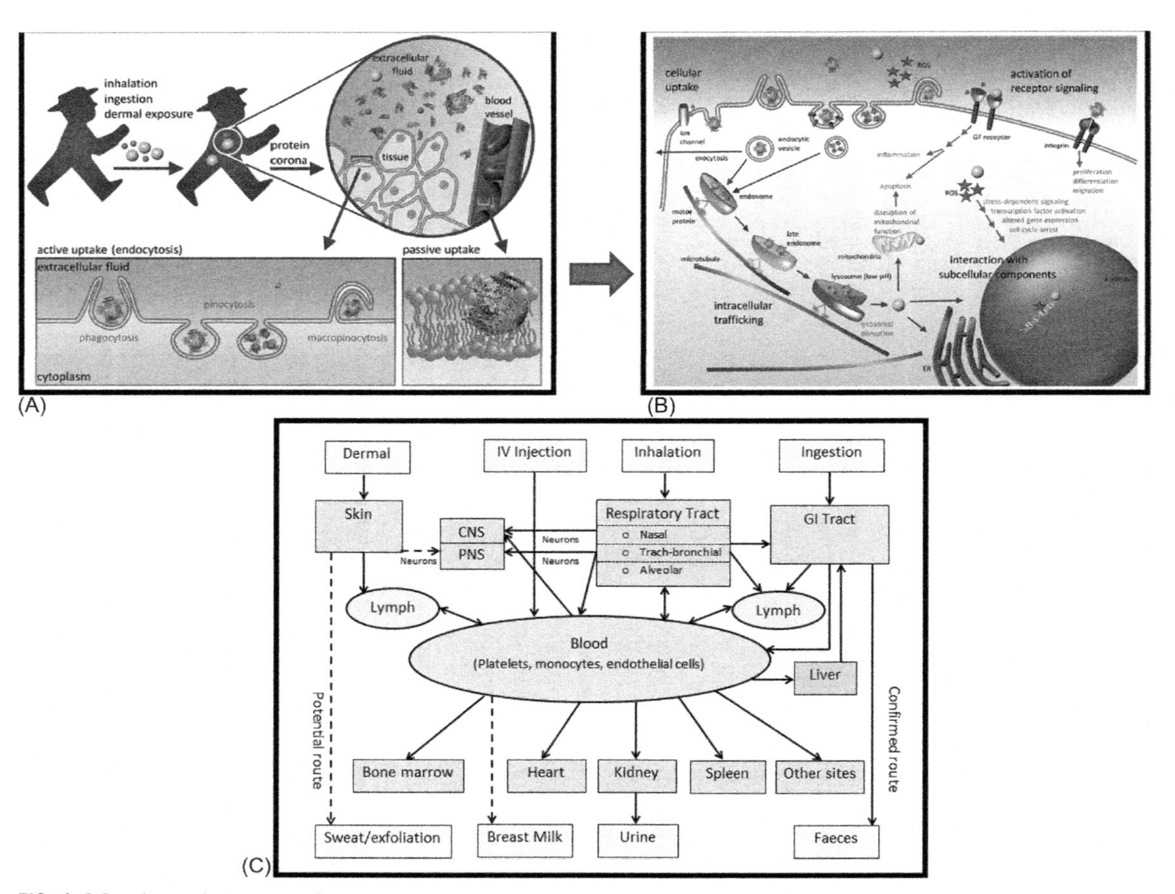

FIG. 1.4 See legends on opposite page

advantage of nanotechnology-based applications, but with complete assurance of health and safety protection? The prediction is that the complete process will take several decades, and even more from a government institutional point of view than from industry. Therefore, the industrial association is a must for a faster process of nanomaterial commercialization. Well-defined strategies must be developed (thoroughly analyzed by academicians and nongovernmental organizations) for deciding on the parameters of toxicity data to be given on commercialization. This is not as simple as it seems, to determine standard parameters governing ecological and human health, due to the high dispersion in toxicological studies.

- *Risk communication for emerging technologies*: There is a high probability that the scientific and technology development communities will not want to release complete toxicity data to the general public, because that will only delay the process of nanomaterial commercialization, as the public would not be accepting of nanomaterials for use in their daily lives once they know the hazards.

FIG. 1.4 CONT'D

(A) Uptake of nanoparticles (NPs): They can enter the human body via inhalation, ingestion, or through the skin. In the extracellular fluid, NPs are coated by proteins and other biomolecules. The so-called protein corona determines how the NP interacts with a cell. Cellular internalization may involve active (receptor-mediated) or passive transport across the cell membrane. (B) Cytotoxic effects of NPs: In the biological environment, NPs can trigger the production of ROS (reactive oxygen species). Elevated ROS levels may lead to (i) activation of cellular stress-dependent signaling pathways, (ii) direct damage of subcellular organelles such as mitochondria, and (iii) DNA fragmentation in the nucleus, resulting in cell cycle arrest, apoptosis, and inflammatory response. NPs may interact with membrane bound to cellular receptors like growth factor (GF) receptors and integrins, including cellular phenotypes such as proliferation, apoptosis, differentiation, and migration. After internalization via endocytic pathways, NPs are trafficked along the endolysosomal network within vesicles with the help of motor proteins and cytoskeletal structures. To access cytoplasmic or nuclear targets, NPs must escape from the endolysosomal network within vesicles with the help of motor proteins and cytoskeletal structures. To access cytoplasmic or nuclear targets, NPs must escape from the endolysosomal network and traverse through the crowded cytoplasm. (C) Biokinetics of NPs: *Solid arrows* (confirmed routes), *dashed arrows* (potential routes); CNS (central nervous system), PNS (peripheral nervous system), GI (gastrointestinal), IV (intravenous).

Courtesy: (A and B) Shang L, Nienhaus K, Nienhaus GU. Engineered nanoparticles interacting with cells: size matters. J Nanobiotechnology 2014;12:1–11. (C) Bakand S, Hayes A. Toxicological considerations, toxicity assessment, and risk management of inhaled nanoparticles. Int J Mol Sci 2016;17:1–17.

Nanotechnology has attempted to solve problems related to various environmental issues and subsequently provide social, economic, and/or political solutions. It has the potential of improving our lives in many different ways. However, incomplete toxicological data related to the environment and human health has created a degree of chaos. Despite heated-up debates on the safety of nanomaterials, some of these materials, particularly semiconductors, have already landed in the commercial marketplace. A thorough analysis of the environmental and health hazards of nanomaterials is a current pressing need, due to the diverse valuable applications being provided by nanotechnology, which cannot be provided by any other known technologies. An open public debate is needed on acceptable technological development, elaborated sustainability objectives, and the degree of environmental impact. However, this is not a simple task, since there is lack of both a research framework and strong leadership. No precise nomenclature has been developed to assess nanotechnology risk analysis and this has created lots of confusion for the public, policymakers, and even researchers, as they are not very clear about the benefits and risks of nanotechnology. Most importantly, the wrong message has been conveyed to the media associated with every class of society. Therefore, the scientific community should do their homework on the precise nomenclature to be used and the risk assessments needed for any nanomaterials before their release into the market. The language should be chosen so that it is understandable to every class of society, in addition to governmental agencies and industries. However, the task of writing such regulations would be quite laborious and very time consuming, requiring many collaborations worldwide to work on various aspects and combine them appropriately. An excellent example of such research coordination is the Global Climate Research Initiative; a similar model could be developed for the case of nanomaterial assessments before their commercialization. Governmental involvement is also required, to closely scrutinize the risks-benefits equation and the level of assessment and regulation to be imposed before nanomaterials can be released for various applications. We are quite far away from making this a reality. There is clear indication of need of supplemental technology that can lessen the negative aspects of nanomaterials while retaining its positive aspects for the solutions of various serious problems.

1.6 Need of combo-technology involving enzyme technology and nanotechnology

As we have seen, nanotechnology has been found to provide solutions for a host of environmental problems, but major concerns exist regarding its human and environmental toxicities. As discussed in the previous sections, there is an urgent need to focus on human health and environmental research before releasing nanotechnology-based products into the commercial market. There should be well-established validated models that can assess the safety of nanomaterials regarding human health and the environment. These models should strongly correlate physical and chemical characteristics of nanomaterials to their behavior and impacts on susceptible populations by developing structure-activity relationships predicting biological impacts, ecological impacts, and degradation at end-of-life. Intensive studies on the physicochemical properties like size, shape, surface charge, composition, coatings, medium, and surface roughness would provide the level of toxicity of any given nanomaterial [156]. Table 1.1 contains the complete summary of correlation of physicochemical properties of nanomaterials with their biological toxicity. The factors that can be used for modifying nanomaterials to nontoxic forms while maintaining their natural characteristics for various applications include: type of medium being used for nanomaterial suspension, surface roughness, and surface coatings.

1. Surface coatings: The adverse effects of nanomaterials can be eliminated by incorporating suitable coatings on their surface. Proper surface coatings could help in stabilizing nanomaterials, avoiding agglomeration, and preventing release of toxic ions. Surface coatings can modify the surface charge and composition of nanomaterials, which can retard their cellular uptake, intracellular distribution, and the production of reactive oxygen species (ROS). The coatings should be chosen so that they are environmentally labile and degrade after being exposed to biological media, thus rendering the nanomaterial nontoxic. It has been shown in various experimental studies that surface coating has been useful in rendering nanomaterials nontoxic [157]. Several studies in animals have found that the level of inflammatory and immunological responses depend on the density and type of coating being used [158]. Generally, biocompatible coatings are preferred due to their easier metabolism. For example, coating with polyethylene glycol (PEG, FDA-approved biocompatible polymer) onto monodisperse Fe_3O_4 nanomaterials

Table 1.1 Correlation Between Physicochemical Properties of Nanomaterials With Biological Toxicity

Nanomaterial Properties	Potential Biological Effects
Size/size distribution (aerodynamic, hydrodynamic)	Crossing tissue and cell membranes, cellular injury, phagocytosis impairment, breakdown in defense mechanisms, migration to other organs, transportation of other environmental pollutants
Surface properties (surface area/mass ratio)	Increased reactivity, increased toxicity
Chemical composition (surface characteristics)	ROS generation, oxidative stress, inflammation, cytokine production, glutathione depletion, mitochondrial exhaustion, cellular injury, protein, and DNA damage
Insolubility or low water solubility	Bioaccumulation inside living systems (such as human cells, tissues and lungs), potential long-term effects
Agglomeration/aggregation	Interruption of cellular processes, cellular injury

produces only negligible aggregation in the cell culture conditions and reduced uptake by macrophage cells. Gold nanomaterials (13-nm) PEGylated with PEG 5000 have been found to be accumulated in mouse liver and spleen for up to 7 days after injection and had longer blood circulation time of about 30h. Studies have found that different molecular weights of PEG can be used for coatings depending on the type of tissue and organ as well as level of exposure to nanomaterials. It has been found that the higher the toxicity of the nanomaterials, the higher would be the layering of the coating to prevent any leaching. There should be proper analysis on the charge of the coating being used for nanomaterials. For example, carboxyl-coated CdSe/ZnS QDs have led to activation of a coagulation cascade leading to pulmonary vascular thrombosis, while coating them with negatively charged serum protein albumin has increased their liver uptake (99%) and faster blood clearance with respect to the latter case. Therefore, surface coatings should be such that they are stable under oxidative or photolytic conditions and have minimal interaction with biological components [159, 160].

2. Surface roughness: The physical surface properties of nanomaterials have been crucial, because they are determinants of cell interactions and subsequently uptake. Generally, surface hydrophobicity and cationic charge are the main factors involved in nonspecific binding of nanomaterials onto the cell and promoting their uptake. It has been found that the smaller the size of the nanomaterials, the higher would be the risk caused by surface roughness due to enhanced electrostatic or hydrophobic/hydrophilic interactions, which promote cell adhesion. Further, they can easily pass through cell membranes by disrupting the phospholipid bilayer of the plasma membrane and create transient holes leading to cell cytotoxicity. It has also been found that a long ordered porous structure present in nanomaterials like silica nanomaterials has led to increased hemolysis due to high penetration through the membrane. Therefore, proper modulation of surface coarseness and porosity could be helpful in reducing the toxicity of nanoparticles [143, 161].
3. Suspending medium: There must be proper and stable dispersion of nanomaterials through a suitable medium, because this is an important factor in determining the degree of agglomeration in nanomaterials. Medium conditions like ionic strength, pH, and temperature have crucial roles in the dispersion of nanomaterials. For example, nanomaterials like TiO_2, ZnO, or carbon black have significantly greater size in PBS (phosphate buffered saline) than in water [162]. Single or clustered nanomaterials have different biological reactivities. Therefore, having a stable suspension of nanomaterials is important before they undergo their toxicity studies in biological systems. It has been recommended to carry out surface modification of nanomaterials through artificial coatings, making them both stable and biocompatible. However, the modification should not affect the physicochemical properties of the nanomaterials.

It has been found that modifying nanomaterials into their nontoxic forms can be done by immobilizing enzymes onto their surface [163]. Enzyme-bound nanomaterials show small Brownian motion, and therefore are less prone to aggregation. Further, the presence of enzyme molecules also minimizes the interaction of nanomaterials with the cell through steric hindrances and leads to lowering of the surface energy. Enzyme molecules are biodegradable and their additional characteristics of catalysis make nanomaterials more versatile and effective in environmental remediation and production of clean energy [11]. On the other hand, enzyme molecules being immobilized onto nanomaterials are less prone to diffusional constraints, highly stable, reusable, and have improved kinetic properties. Enzymes being immobilized onto nanomaterials makes them resistant to unfolding and gives enhanced stability in extreme physicochemical parameters, with easier handling and reduction in total cost of the process by severalfold. Nanomaterials have very large surface area, leading to very high enzyme loading, thus improving immobilization efficiency [164]. Further, enzyme purity is not as important when immobilizing enzymes onto nanomaterials, because there is a tremendous increment in specific activity of the enzyme during immobilization. There is easier separation of immobilized enzymes from the reaction mixture, particularly when magnetic nanomaterials are used as the immobilizing matrix. Immobilized enzymes onto nanomaterials can be easily packed into any type of reactor with varied configuration due to having limiting size constraints. Even multimeric enzymes like oxidoreductases can be stabilized by their immobilization onto nanomaterials. As enzymes are immobilized onto the surface of nanomaterials, they exhibit better activity than those confined into cavities. It has been found that enzyme immobilization leads to structural changes, particularly gaining β-sheet structure and losing the α-helical structure on various solid substrates. However, no such changes are found when enzymes are being immobilized onto nanomaterials and they maintain their native structures [165]. Enzymes can be linked to nanomaterials in the following ways [10]:

- *Electrostatic adsorption*: It is the simplest and most widely used approach in which enzymes are attached onto nanomaterials through ionic interactions. During interaction, the pH and ionic strength of the medium is appropriately modulated to obtain strong interaction between the enzyme and the nanomaterials.
- *Covalent attachment*: Enzymes are covalently linked to nanomaterials via functional groups. Nanomaterials are modified with a variety of organic functional groups using mild conditions. The popular labeling chemistry utilizes the covalent binding of primary amines with sulfo-NHS (N-hydroxysuccinimide) esters or R-COOH groups via reaction with EDC (1-ethyl-3-(3-dimethylaminopropyl) carbodiimide hydrochloride). Nanomaterials labeled with NHS esters can react to form covalent bonds with the primary amine of lysine present on the enzyme. In some cases, nanomaterials are coated with maleimide groups that are used to react with the thiol group of cysteine present on the enzyme. The nanomaterial oxides like TiO_2, Fe_2O_3, CuO, AgO, and AuO are generally modified by silanization, which yields a modified surface exhibiting amino groups used for coupling linking enzymes.

- *Conjugation via specific affinity*: Enzymes prepared by recombinant technology are added with specific groups, which help in their attachment onto nanomaterials. For example, nanomaterials coated with streptavidin are used to selectively bind biotin-labeled enzymes.
- *Direct conjugation to the nanoparticle surface*: Enzymes are directly attached onto nanomaterials without any use of a linker through direct reaction of a chemical group present on the enzyme: for example, attachment of enzymes onto Au and Ag nanoparticles through thiol groups of cysteine forming covalent bonds with the nanoparticles. In sulfur-containing nanomaterials like ZnS/CdSe, cysteine forms a direct disulfide bridge with the surface S atom. Direct linkages can also be achieved in enzymes having His tags, which help in direct attachment of enzymes to nanoparticles like Zn, Ni, Cu, Co, Fe, Mn, etc. Generally direct linkages are used when enzymes immobilized onto nanomaterials are implemented for biosensors in which FRET (Förster resonance energy transfer) or electron transfer is used.
- *Magnetic interaction*: Magnetic nanomaterials are used for enzyme inmmobilization as they favor high binding capacity and high catalytic specificity to the conjugated enzyme. More significantly, they help in efficient recovery of the enzyme without contaminating the final product. Enzyme immobilization onto magnetic nanomaterials could either be direct or through modification of magnetic nanomaterials. It has been found that an enzyme being immobilized onto nickel-impregnated silica paramagnetic particles (NSP) has improved catalytic properties with respect to its soluble counterpart as well as NSP protecting the enzyme from deactivation at extreme pH [166].

Nanotechnology and enzyme technology complement each other and the combination of the two technologies could be helpful for commercialization and could pose a lower risk in large-scale or small-scale industries. Studies are in progress to show the complete nontoxicity of nanomaterials with enzymes immobilized on them. The following chapters of this book discuss immobilized enzymes onto nanomaterials used for environmental remediation and production of clean energy using green techniques. Some of the applications have been successfully patented and are in process to be placed on the market. Finally, the sustainability of combining the two technologies has been thoroughly discussed.

References

[1] Wang F. ChemSusChem 2017;10:4393–402.

[2] Novikov LS, Voronina EN. Potential space applications of nanomaterials. New York: Springer Nature; 2017, p. 139–47.

[3] Moon RJ, Martini A, Nairn J, Simonsen J, Youngblood J. Chem Soc Rev 2011;40:3941–94.

[4] www.bccresearch.com/market-research/nanotechnology.

[5] Le TS, Dao TH, Nguyen DC, Nguyen HC, Balikhin IL. Adv Nat Sci Nanosci Nanotechnol 2015;6:1–8.

[6] Lin J, Shang Y, Ding B, Yang J, Yu J, SS A-D. Mar Pollut Bull 2012;64:347–52.
[7] Zhang R, Liu C, Hsu P-C, Zhang C, Liu N, Zhang J, Lee HR, Lu Y, Qiu Y, Chu S, Cui Y. Nano Lett 2016;16:3642–9.
[8] Hristozov D, Ertel J. Forum Forsch 2009;22:161–8.
[9] Bowman DM, Hodge GA. Futures 2006;38:1060–73.
[10] Dwevedi A. Enzyme immobilization: advancement in industry, agriculture, medicine and environment. New York: Springer Nature Publishers; 2016, p. 21–44.
[11] Alcalde M, Ferrer M, Plou FJ, Ballesteros A. Trends Biotechnol 2006;24:281–7.
[12] Wait AF, Parkin A, Morley GM, dos Santos L, Armstrong FA. J Phys Chem C 2010;114:12003–9.
[13] Bansal A, Illukpitiya P, Tegegne F, Singh SP. Renew Sust Energ Rev 2016;58:141–6.
[14] Chen M, Zeng G, Xu P, Lai C, Tang L. Trends Biochem Sci 2017;42:914–30.
[15] Adachi M, Murata Y, Takao J, Jiu J, Sakamoto M, Wang F. J Am Chem Soc 2004;126:14943–9.
[16] Law M, Greene LE, Johnson JC, Saykally R, Yang P. Nat Mater 2005;4:455–9.
[17] Green AN, Palomares E, Haque SA, Kroon JM, Durrant JR. J Phys Chem B 2005;109:12525–33.
[18] Zhou L, Sun Y, Yang Z, Zhou Y. J Colloid Interface Sci 2005;289:347–51.
[19] Patchkovskii S, Tse JS, Yurchenko SN, Zhechkov L, Heine T, Seifert G. Proc Natl Acad Sci U S A 2005;102:10439–44.
[20] Chen J, Cheng F. Acc Chem Res 2009;42:713–23.
[21] Liu HK, Wang GX, Guo Z, Wang J, Konstantinov K. J Nanosci Nanotechnol 2006;6:1–15.
[22] Tenne R. Nat Nanotechnol 2006;1:103–11.
[23] Umeyama T, Imahori H. Photosynth Res 2006;87:63–71.
[24] Lux KW, Rodriguez KJ. Nano Lett 2006;6:288–95.
[25] Li S, Zhu B. J Nanosci Nanotechnol 2009;9:3824–7.
[26] Chu D, Wang S, Zheng P, Wang J, Zha L, Hou Y, He J, Xiao Y, Lin H, Tian Z. ChemSusChem 2009;2:171–6.
[27] Farhat S, Weinberger B, Lamari FD, Izouyar T, Noe L, Monthioux M. J Nanosci Nanotechnol 2007;7:3537–42.
[28] Kowalczyk P, Hołyst R, Terrones M, Terrones H. Phys Chem Chem Phys 2007;9:1786–92.
[29] Alayoglu S, Nilekar AU, Mavrikakis M, Eichhorn B. Nat Mater 2008;7:333–8.
[30] Toma FM, Sartorel A, Iurlo M, Carraro M, Parisse P, Maccato C, Rapino S, Gonzalez BR, Amenitsch H, Da Ros T, Casalis L, Goldoni A, Marcaccio M, Scorrano G, Scoles G, Paolucci F, Prato M, Bonchio M. Nat Chem 2010;2:826–31.
[31] Hasobe T, Fukuzumi S, Kamat PV. J Phys Chem B 2006;110:25477–84.
[32] Yang MD, Liu YK, Shen JL, Wu CH, Lin CA, Chang WH, Wang HH, Yeh HI, Chan WH, Parak WJ. Opt Express 2008;16:15754–8.
[33] Yang N, Zhai J, Wang D, Chen Y, Jiang L. ACS Nano 2010;4:887–94.
[34] Perlich J, Kaune G, Memesa M, Gutmann JS, Müller-Buschbaum P. Philos Transact A Math Phys Eng Sci 2009;367:1783–98.
[35] Kongkanand A, Tvrdy K, Takechi K, Kuno M, Kamat PV. J Am Chem Soc 2008;130:4007–15.
[36] Xiang JH, Zhu PX, Masuda Y, Okuya M, Kaneko S, Koumoto K. J Nanosci Nanotechnol 2006;6:1797–801.
[37] Umar A. Nanoscale Res Lett 2009;4:1004–8.
[38] Gnichwitz JF, Marczak R, Werner F, Lang N, Jux N, Guldi DM, Peukert W, Hirsch A. J Am Chem Soc 2010;132:17910–20.

[39] Tian B, Zheng X, Kempa TJ, Fang Y, Yu N, Yu G, Huang J, Lieber CM. Nature 2007;449:885–9.
[40] Mutlugun E, Soganci IM, Demir HV. Opt Express 2008;16:3537–45.
[41] Liu CY, Holman ZC, Kortshagen UR. Nano Lett 2009;9:449–52.
[42] Ossicini S, Amato M, Guerra R, Palummo M, Pulci O. Nanoscale Res Lett 2010;5:1637–49.
[43] Klinger C, Patel Y, Postma HWC. PLoS One 2012;7:e37806.
[44] Kim J, Hong AJ, Chandra B, Tulevski GS, Sadana DK. Adv Mater 2012;24:1899–902.
[45] Bernardi M, Lohrman J, Kumar PV, Kirkeminde A, Ferralis N, Grossman JC, Ren S. ACS Nano 2012;6:8896–903.
[46] Guo W, Chen C, Ye M, Lv M, Lin C. Nanoscale 2014;6:3656–63.
[47] Ganapathy V, Kong EH, Park YC, Jang HM, Rhee SW. Nanoscale 2014;6:3296–301.
[48] Zhang W, Gu J, Liu Q, Su H, Fan T, Zhang D. Phys Chem Chem Phys 2014;16:19767–80.
[49] Liu Y, Qi N, Song T, Jia M, Xia Z, Yuan Z, Yuan W, Zhang KQ, Sun B. ACS Appl Mater Interfaces 2014;6:20670–5.
[50] Qiu F, Li Y, Yang D, Li X, Sun P. Bioresour Technol 2011;102:4150–6.
[51] Feyzi M, Norouzi L, Rafiee HR. Sci World J 2013;2013:612712.
[52] Wang YT, Fang Z, Zhang F, Xue BJ. Bioresour Technol 2015;193:84–9.
[53] Sajeevan AC, Sajith V. J Eng 2013;2013:1–9.
[54] Wu C, Wang Z, Williams PT, Huang J. J Sci Rep 2013;3:2742.
[55] Filanovsky B, Granot E, Dirawi R, Presman I, Kuras I, Patolsky F. Nano Lett 2011;11:1727–32.
[56] Vinayan BP, Ramaprabhu S. Nanoscale 2013;5:5109–18.
[57] Maya-Cornejo J, Ortiz-Ortega E, Álvarez-Contreras L, Arjona N, Guerra-Balcázar M, Ledesma-García J, Arriaga LG. Chem Commun (Camb) 2015;51:2536–9.
[58] Lu Y, Jin R, Chen W. Nanoscale 2011;3:2476–80.
[59] Tozzini V, Pellegrini V. Phys Chem Chem Phys 2013;15:80–9.
[60] Yeh TF, Teng CY, Chen SJ, Teng H. Adv Mater 2014;26:3297–303.
[61] Haider Z, Kang YS. ACS Appl Mater Interfaces 2014;6:10342–52.
[62] Morozan A, Donck S, Artero V, Gravel E, Doris E. Nanoscale 2015;7:17274–7.
[63] Xu Y, Yin X, Huang Y, Du P, Zhang B. Chemistry 2015;21:4571–5.
[64] Xin S, Guo YG, Wan LJ. Acc Chem Res 2012;45:1759–69.
[65] Sassin MB, Chervin CN, Rolison DR, Long JW. Acc Chem Res 2013;46:1062–74.
[66] Rui X, Tan H, Yan Q. Nanoscale 2014;6:9889–924.
[67] Tang X, Yan F, Wei Y, Zhang M, Wang T, Zhang T. ACS Appl Mater Interfaces 2015;7:21890–7.
[68] Merugula L, Khanna V, Bakshi BR. Environ Sci Technol 2012;46:9785–92.
[69] Ma P-C, Zhang Y. Renew Sust Energ Rev 2014;30:651–60.
[70] Li Y, Lu J. Mater Des 2014;57:689–96.
[71] Ng K-W, Lam W-H, Pichiah S. Renew Sust Energ Rev 2013;28:331–9.
[72] Wang H, Lin J, Shen ZX. J Sci Adv Mat Dev 2016;1:225–55.
[73] van Dijk L, Paetzold UW, Blab GA, Schropp RE, di Vece M. Prog Photovolt 2016;24:623–33.
[74] Zhang L, Wang ZS. Chem Asian J 2016;11:3283–9.
[75] Zhong L, Yu F, An Y, Zhao Y, Sun Y, Li Z, Lin T, Lin Y, Qi X, Dai Y, Gu L, Hu J, Jin S, Shen Q, Wang H. Nature 2016;538:84–7.
[76] Chaturvedi S, Dave PN, Shah NK. J Saudi Chem Soc 2012;16:307–25.

[77] Hoa LQ, Vestergaard MC, Tamiya E. Sensors (Basel) 2017;17:2587–608.
[78] Aiso K, Takeuchi R, Masaki T, Chandra D, Saito K, Yui T, Yagi M. ChemSusChem 2017;10:687–92.
[79] Bang JJ, Guerrero PA, Lopez DA, Murr LE, Esquivel EV. J Nanosci Nanotechnol 2004;4:716–8.
[80] Tungittiplakorn W, Cohen C, Lion LW. Environ Sci Technol 2005;39:1354–8.
[81] Liu Y, Chen X, Li J, Burda C. Chemosphere 2005;61:11–8.
[82] Lu C, Chung YL, Chang KF. Water Res 2005;39:1183–9.
[83] Dror I, Baram D, Berkowitz B. Environ Sci Technol 2005;39:1283–90.
[84] Liu Y, Majetich SA, Tilton RD, Sholl DS, Lowry GV. Environ Sci Technol 2005;39:1338–45.
[85] Lin CJ, Lo SL, Liou YH. Chemosphere 2005;59:1299–307.
[86] He F, Zhao D. Environ Sci Technol 2005;39:3314–20.
[87] Hu J, Chen G, Lo IM. Water Res 2005;39:4528–36.
[88] Nair AS, Pradeep T. J Nanosci Nanotechnol 2007;7:1871–7.
[89] Rizzo L, Uyguner CS, Selcuk H, Bekbolet M, Anderson M. Water Sci Technol 2007;55:113–8.
[90] Dai K, Chen H, Peng T, Ke D, Yi H. Chemosphere 2007;69:1361–7.
[91] Coleman HM, Vimonses V, Leslie G, Amal R. J Hazard Mater 2007;146:496–501.
[92] Chiou CH, Juang RS. J Hazard Mater 2007;149:1–7.
[93] Jézéquel H, Chu KH. J Environ Sci Health A Tox Hazard Subst Environ Eng 2006;41:1519–28.
[94] Liu JF, Zhao ZS, Jiang GB. Environ Sci Technol 2008;42:6949–54.
[95] Chen SY, Chen WH, Shih CJ. Water Sci Technol 2008;58:1947–54.
[96] Tawabini B, Al-Khaldi S, Atieh M, Khaled M. Water Sci Technol 2010;61:591–8.
[97] Varanasi P, Fullana A, Sidhu S. Chemosphere 2007;66:1031–8.
[98] Uyak V, Koyuncu I, Oktem I, Cakmakci M, Toroz I. J Hazard Mater 2008;152:789–94.
[99] Zhu N, Luan H, Yuan S, Chen J, Wu X, Wang L. J Hazard Mater 2010;176:1101–15.
[100] Song Z, Chen L, Hu J, Richards R. Nanotechnology 2009;20:275707.
[101] Moussavi G, Mahmoudi M. J Hazard Mater 2009;168:806–12.
[102] Wazne M, Meng X, Korfiatis GP, Christodoulatos C. J Hazard Mater 2006;136:47–52.
[103] Urban I, Ratcliffe NM, Duffield JR, Elder GR, Patton D. Chem Commun (Camb) 2010;46:4583–5.
[104] Tong M, Yuan S, Long H, Zheng M, Wang L, Chen J. J Contam Hydrol 2011;122:16–25.
[105] Wu D, Shen Y, Ding A, Qiu M, Yang Q, Zheng S. Environ Technol 2013;34:2663–9.
[106] Liu H, Guo M, Zhang Y. Environ Technol 2014;35:917–24.
[107] Tofighy MA, Mohammadi T. J Hazard Mater 2011;185:140–7.
[108] Nabid MR, Sedghi R, Behbahani M, Arvan B, Heravi MM, Oskooie HA. J Mol Recognit 2014;27:421–8.
[109] Suman, Kardam A, Gera M, Jain VK. Environ Technol 2015;36:706–14.
[110] Chu W, Zhang Y, Li D, Barrow CJ, Wang H, Yang W. Biosens Bioelectron 2015;67:621–4.
[111] Yang D, Liu H, Zheng Z, Sarina S, Zhu H. Nanoscale 2013;5:2232–42.
[112] Lin S, Huang R, Cheng Y, Liu J, Lau BL, Wiesner MR. Water Res 2013;47:3959–65.
[113] Lubasova D, Netravali A, Parker J, Ingel B. J Nanosci Nanotechnol 2014;14:4891–8.
[114] Celik E, Park H, Choi H, Choi H. Water Res 2011;45:274–82.
[115] Rasheed QJ, Pandian K, Muthukumar K. Ultrason Sonochem 2011;18:1138–42.
[116] Lin X, Lu F, Chen Y, Liu N, Cao Y, Xu L, Wei Y, Feng L. ACS Appl Mater Interfaces 2015;7:8108–13.

[117] Gethard K, Sae-Khow O, Mitra S. ACS Appl Mater Interfaces 2011;3:110–4.
[118] Chua YT, Lin CX, Kleitz F, Zhao XS, Smart S. Chem Commun (Camb) 2013;49:4534–6.
[119] Ruiz MP, Faria J, Shen M, Drexler S, Prasomsri T, Resasco DE. ChemSusChem 2011;4:964–74.
[120] Chen X, Hong L, Xu Y, Ong ZW. ACS Appl Mater Interfaces 2012;4:1909–18.
[121] Yi XS, Yu SL, Shi WX, Wang S, Jin LM, Sun N, Ma C, Sun LP. Water Sci Technol 2013;67:477–84.
[122] Iqbal MZ, Abdala AA. Environ Sci Pollut Res Int 2013;20:3271–9.
[123] He Y, Liu Y, Wu T, Ma J, Wang X, Gong Q, Kong W, Xing F, Liu Y, Gao J. J Hazard Mater 2013;260:796–805.
[124] Yang S, Chen L, Mu L, Ma PC. J Colloid Interface Sci 2014;430:337–44.
[125] Kim DH, Jung MC, Cho S-H, Kim SH, Kim H-Y, Lee HJ, Oh KH, Moon M-W. Sci Rep 2015;5:12908.
[126] Sim KM, Park HS, Bae GN, Jung JH. Sci Total Environ 2015;533:266–74.
[127] Li X, Wang N, Fan G, Yu J, Gao J, Sun G, Ding B. J Colloid Interface Sci 2015;439:12–20.
[128] Saraee HS, Jafarmadar S, Taghavifar H, Ashrafi SJ. Int J Environ Sci Technol 2015;12:2245–52.
[129] Liu W, Ma J, Shen C, Wenn Y, Liu W. Water Res 2016;90:24–33.
[130] Peter KT, Vargo JD, Rupasinghe TP, Jesus AD, Tivanski AV, Sander EA, Myung NV, Cwiertny DM. ACS Appl Mater Interfaces 2016;8:11431–40.
[131] Hena S, Rozi R, Tabassum S, Huda A. Environ Sci Pollut Res Int 2016;23:14868–80.
[132] Peter KT, Johns AJ, Myung NV, Cwiertny DM. Water Res 2017;117:207–17.
[133] Deng Q, Pa J, Yin X, Wang X, Zhao L, Kang SG, Jimenez-Cruz CA, Zhou R, Li J. Phys Chem Chem Phys 2016;18:8140–7.
[134] Nthumbi RM, Ngila JC. Environ Sci Pollut Res Int 2016;23:20214–31.
[135] Jing L, Shim K, Toe CY, Fang T, Zhao C, Amal R, Sun KN, Kim JH, Ng YH. ACS Appl Mater Interfaces 2016;8:7030–6.
[136] Zhang B, Zhang ZG, Yan X, Wang XX, Zhao H, Guo J, Feng JY, Long YZ. Nanoscale 2017;9:4154–61.
[137] Wongaree M, Chiarakorn S, Chuangchote S, Sagawa T. Environ Sci Pollut Res Int 2016;23:21395–406.
[138] Dehghan SF, Golbabaei F, Maddah B, Latifi M, Pezeshk H, Hasanzadeh M, Akbar-Khanzadeh F. J Air Waste Manage Assoc 2016;66:912–21.
[139] Wang Q, Hung PC, Lu S, Chang MB. Chemosphere 2016;159:132–7.
[140] Gehrke I, Geiser A, Somborn-Schulz A. Nanotechnol Sci Appl 2015;8:1–17.
[141] Kessler R. Environ Health Perspect 2011;119:17.
[142] Ray PC, Yu H, Fu PP. J Environ Sci Health C Environ Carcinog Ecotoxicol Rev 2009;27:1–35.
[143] Sharifi S, Behzadi S, Laurent S, Forrest ML, Stroeve P, Mahmoudi M. Chem Soc Rev 2012;41:2323–43.
[144] Hurt RH, Monthioux M, Kane A. Carbon 2006;44:1028–33.
[145] Huang Y-W, Wu C-H, Aronstam RS. Materials (Basel) 2010;3:4842–59.
[146] Song Y, Tang S. Sci World J 2011;11:1821–8.
[147] Shang L, Nienhaus K, Nienhaus GU. J Nanotechnol 2014;12:1–11.
[148] Sharma HS, Hussain S, Schlager J, Ali SF, Sharma A. Acta Neurochir Suppl 2010;106:359–64.

[149] Nielsen GD, Roursgaard M, Jensen KA, Poulsen SS, Larsen ST. Basic Clin Pharmacol Toxicol 2008;103:197–208.
[150] Shi H, Magaye R, Castranova V, Zhao J. Part Fibre Toxicol 2013;10:1–33.
[151] Yang H, Liu C, Yang D, Zhang H, Xi Z. J Appl Toxicol 2009;29:69–78.
[152] Feng X, Chen A, Zhang Y, Wang J, Shao L, Wei L. Int J Nanomedicine 2015;10:4321–40.
[153] Bakand S, Hayes A. Int J Mol Sci 2016;17:929–46.
[154] Handy RD, Shaw BJ. Health Risk Soc 2007;9:125–44.
[155] Fu PP, Xia Q, Hwang H-M, Ray PC, Yu H. J Food Drug Anal 2014;22:64–75.
[156] Aillon KL, Xie Y, El-Gendy N, Berkland CJ, Forrest ML. Adv Drug Deliv Rev 2009;61:457–66.
[157] Gatoo MA, Naseem S, Arfat MY, Dar AM, Qasim K, Zubair S. Biomed Res Int 2014;2014:1–8.
[158] Ngobili TA, Daniele MA. Exp Biol Med (Maywood) 2016;241:1064–73.
[159] Baer DR, Engelhard MH, Johnson GE, Laskin J, Lai J, Mueller K, Munusamy P, Thevuthasan S, Wang H, Washton N. J Vac Sci Technol A 2013;31:1–32.
[160] Baer DR. J Surf Anal 2011;17:163–9.
[161] Kim ST, Saha K, Kim C, Rotello VM. Acc Chem Res 2013;46:681–91.
[162] Pavlina M, Bregar VB. Dig J Nanomater Biostruct 2012;7:1389–400.
[163] Verma ML, Puri M, Barrow CJ. Crit Rev Biotechnol 2016;36:108–19.
[164] Ding S, Cargill AA, Medintz IL, Claussen JC. Curr Opin Biotechnol 2015;34:242–50.
[165] Secundo F. Chem Soc Rev 2013;42:6250–61.
[166] Johnson PA, Park HJ, Driscoll AJ. Methods Mol Biol 2011;679:183–91.

CHAPTER

Production of clean energy by green ways

2

Dinesh Pratap Singh*[,a], Alka Dwevedi[†,‡,a]

Department of Physics, University of Santiago, Estacion Central Santiago, Chile Department of Biotechnology, Delhi Technological University, New Delhi, India[†] Swami Shraddhanand College, University of Delhi, New Delhi, India[‡]*

2.1 Introduction

Energy is the most basic requirement for the existence of life on earth. It is an integral part of daily life, from our rising in the morning to going to bed at night. Energy requirements are increasing day by day due to improvements in the quality of life. All sectors are dependent on energy, from industry and transportation to the myriad comforts and conveniences of the home and workplace, and even national security. Our appetite for energy has become boundless. However, energy sources are circumscribed (~85% of total energy is derived from fossil fuels like oil, coal, and gas), as usage of fossil fuels occurs much faster than they are formed. Development has been directly correlated with increased use of fossil fuels (source of energy), with worldwide demand estimated to double by 2030.

Various reports have been published on understanding the link between economic growth and energy consumption. The energy consumption related to commercial and industrial applications has an effective impact on every single citizen through the cost of goods and services, quality of manufactured products, economical strength, and job availability. In one recent report, energy consumption was related to economic growth, represented by gross domestic product (GDP) per capita, with statistically significant results, showing that there is a 0.82% increase in GDP when energy consumption has increased by 1% in the case of upper middle income countries, 0.81% increase in GDP when energy consumption has increased by 1% in the case of lower middle income countries, while a 0.73% increase in GDP has been found when energy consumption increases by 1% in the case of lower income countries [1]. The results imply that upper middle– and lower middle–income countries are more energy dependent than lower income countries. However, no literature has found a direct correlation between the energy consumption and GDP growth nexus, due to the wide range of differences in results between developed and developing countries,

[a]Both authors contributed equally.

Solutions to Environmental Problems Involving Nanotechnology and Enzyme Technology
https://doi.org/10.1016/B978-0-12-813123-7.00002-5

as well as resource-endowed and non–resource endowed countries. Interestingly, a study conducted by Climate Institute (2013) has found that there is positive correlation between energy efficiency and economic growth based on the data obtained from 28 diverse economies over a period of three decades. It has been reported that there is strong correlation between energy efficiency and economic growth. It has been found that a 1% increase in level of energy efficiency has resulted in a 0.1% increment in the rate of economic growth. This clearly indicates that it is not energy source that is important in determining economic growth, but the amount of energy produced from any source is actually the determining factor.

A variety of energy sources are available on earth, categorized as renewable (sun, wind, water, etc.) and nonrenewable sources (fossil fuels, radioactive elements, etc.). Nonrenewable sources of energy are very limited; however, demand is increasing rapidly due to rapid industrialization and improving daily lifestyles. The immense usage of fossil fuels has created an alarming situation with respect to the amount as well as the release of toxic gases like CO_2, CO, SO_2, NO, NO_2, particulate matter, and heavy metals. These toxic substances are harmful to the environment and human health, as well as being related to global warming and ozone depletion [2, 3] (Fig. 2.1). Rate of consumption of various known energy sources like oil, coal and radioactive elements are reported as 40%, 23%, and 8%, respectively Coal is the cheapest fossil fuel but it is the major source of CO_2 (greenhouse gas) emissions due to burning and mining. The CO_2 concentration in the atmosphere has risen by about 40%, i.e. from 270 ppm to 380 ppm, since the beginning of the Industrial Revolution. In case of United States, total greenhouse gas emissions have been found to be 6870 million metric tons in 2014 (USEPA). It will continue to rise if no significant step will be taken soon. It has been found that total CO_2 emissions were 5.9 billion metrics ton in 2006 and are estimated to be 7.4 billion metric tons in 2030.

Nuclear energy has been found to be clean to some extent, with generation of large amounts of energy within a very short time span. However, it is the most hazardous, with various examples of disasters like the release of radioactive materials at the Fukushima I Nuclear Power Plant following the earthquake and tsunami on 11 March 2011 [4, 5]. Some countries are discontinuing nuclear energy. Germany formally announced on May 30, 2011 that it would completely abandon nuclear energy within the coming 11 years [6]. Therefore alternate sources of energy that are long lasting, clean (no generation of toxic gases), cheap, and easily accessible must be found.

Worldwide consumption of fossil fuels has put increased pressure on traditional energy sources, particularly renewable sources. Renewable energy sources can provide long-term energy availability, increase diversity of energy sources, promote regional development (since they can be used even in undeveloped areas without conventional energy sources), and effectively reduce the cost associated with climate change. Both technical and financial interests favor the utilization of renewable sources with minimal CO_2 emissions. Unlimited energy sources in the form of solar, wind, and hydrogen are available that can be used effectively through suitable technology. Presently, only about 7% of worldwide energy needs are met by renewable

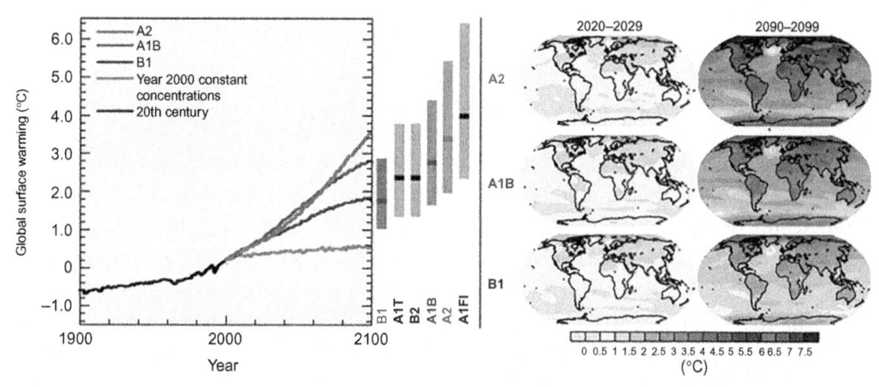

(A)

Indicators of a warming world

Humidity

Glaciers

Temperature over land

Temperature over oceans

Snow cover

Air temperature near surface (troposphere)

Tree lines shifting poleward and upward

Sea surface temperature

Sea level

Spring coming earlier

Sea ice

Ice sheets

Species migrating poleward and upward

Ocean heat content

(B)

FIG. 2.1

Burning of fossil fuels has led to an increase in surface temperature of the planet. (A) Left panel: Solid lines are multimodel global averages of surface warming (1980–1999) for the SRES (IPCC Special Report on Emissions Scenarios) scenarios A2, A1B and B1, shown as continuations of the 20th century simulations. The orange line shows the experimental results when concentrations were held constant at year 2000 values. The bars in the middle of the figure indicate the best estimate (solid line within each bar) and the likely range assessed for the six SRES marker scenarios for 2090–2099 relative to 1980–1999. The assessment of the best estimate and likely ranges in the bars include the Atmosphere–Ocean General Circulation Models (AOGCMs) in the left part of the figure, as well as results from a hierarchy of independent models and observational constraints. Right panels: Projected surface temperature changes for the early and late 21st century relative to the period 1980–1999. The panels show the multi-AOGCM average projections for the A2 (top), A1B (middle) and B1 (bottom) SRES scenarios averaged over the decades 2020–2029 (left) and 2090–2099 (right). (Courtesy: Pachauri, R.K., Reisinger, A. (Eds.) (Core Writing Team). Climate Change 2007: Synthesis Report. Contribution of Working Groups I, II and II to the Fourth Assessment Report of the Intergovernmental Panel on Climate Change. IPCC, Geneva, Switzerland.) (B) Various indicators of global warming leading to increase in surface temperature of earth.

Adapted from: https://skepticalscience.com/graphics.php?g=8.

energy sources, mostly from hydropower and biomass energy (organic matter such as wood, municipal waste, and agricultural crops). It has been estimated that usage of renewable energy sources will increase by over 60% in the next two decades due to their significant contributions towards the energy supply. It has been reported in China that in the decade from 1978 to 2008, there was an increase in GDP and GDP per capita by 0.120% and 0.162%, respectively, with a 1% increase in consumption of renewable energy sources [7] (Fig. 2.2). However, energy price and availability are not the sole determinants in setting economic trends but other factors are involved like laws and regulations governing energy choices, worldwide demand, policies and political agendas, lifestyle choices, business decisions, climate change, and the pace of developments in science and engineering. Quests for technologies to capture maximum energy from renewable sources that are both cost effective and easily accessible are in progress. In this chapter, nanotechnology and enzyme technology are discussed for their implications in capturing energy from renewable sources in a cost effective way.

2.2 Need for clean energy

Increased demand for energy in almost every sector of life has brought society into a situation of energy crisis and severe pollution, due to the generation of various toxic gases in addition to the greenhouse gases responsible for global warming. There is an urgent need for energy sources that are clean, i.e., that don't produce any by-products toxic to human health and the environment. Renewable sources are best suited for providing clean energy, which is better for our energy security, environmental safety, climate change mitigation, and from an economic point of view [8]. Renewable sources can supply unlimited energy through implementation of effective technologies. Factors like cost benefits, stability, efficiency, cleanliness, and environmental friendliness add icing on the cake for renewable sources. Thus, most countries of the world are trying to turn to renewable sources for their energy demands (Fig. 2.3).

Fossil fuels are limited and unsustainable and will eventually be depleted. According to the Energy Information Administration (EIA), OPEC member countries are responsible for ~40% of the world's total oil production and hold the majority of the world's oil reserves. Limited supply, increasing demand, and ~85% dependency on fossil fuels have resulted in drastic increments in worldwide prices of fossil fuels. As previously discussed, fossil fuels have posed a major threat to ecological balance, being a causative agent of many ecological hazards. All fossil fuels (coal, natural gas, and oil) are basically compounds of hydrocarbons. Therefore they produce large amounts of CO_2 on burning, which is the perpetrator of global warming. Global warming has increased the temperature of our planet, leading to ice melting in the Arctic and Antarctica, causing higher sea levels than normal. Further, higher temperatures have also led to endangering of species (both flora and fauna) living both on land and in the sea, in addition to increased flooding, which has hurt agricultural and fishing activities. Other toxic gases like NO, NO_2, SO_2, CO, and

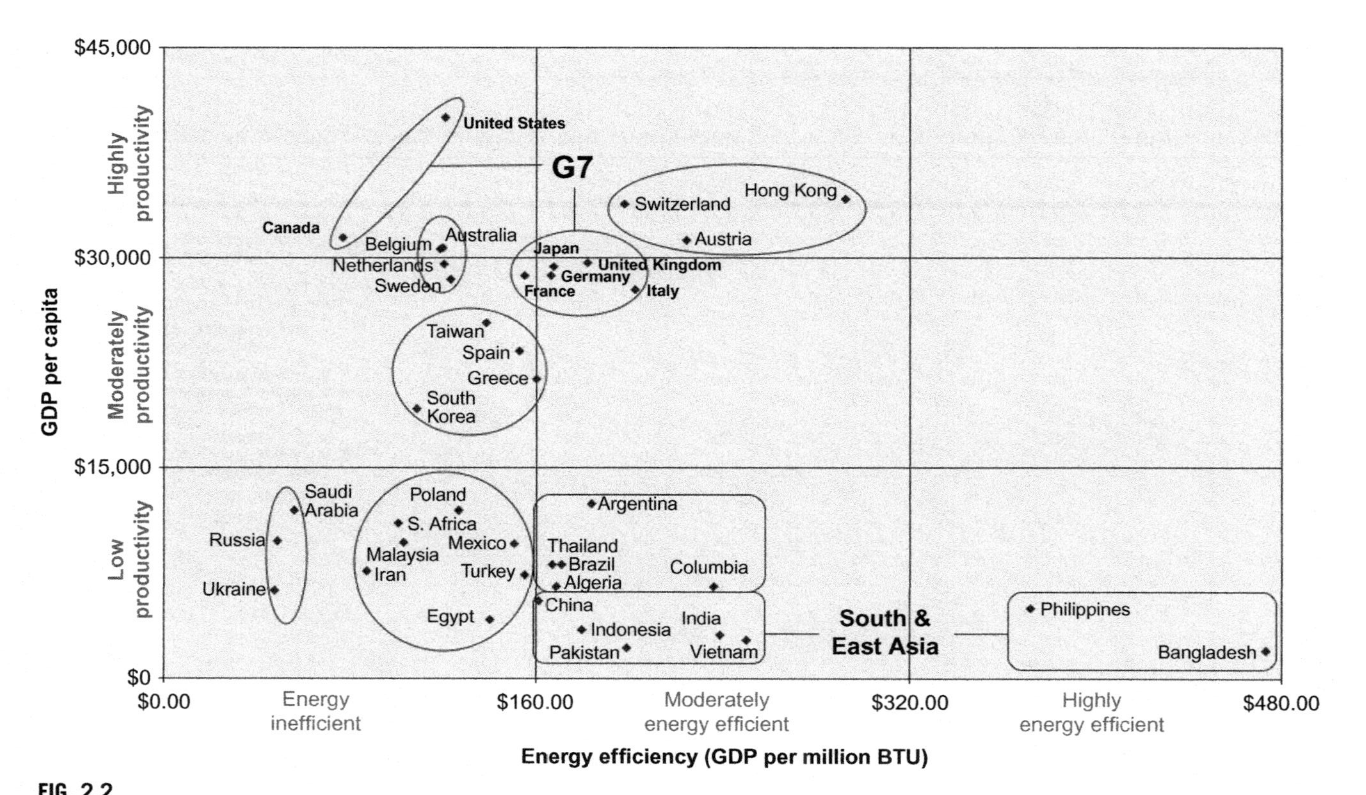

FIG. 2.2

Correlation between GDP (gross domestic product) per capita vs. energy efficiency (in BTUs, British Thermal Units) of top 40 economies in the world.

(Adapted from https://www.e-education.psu.edu/earth104/node/999.)

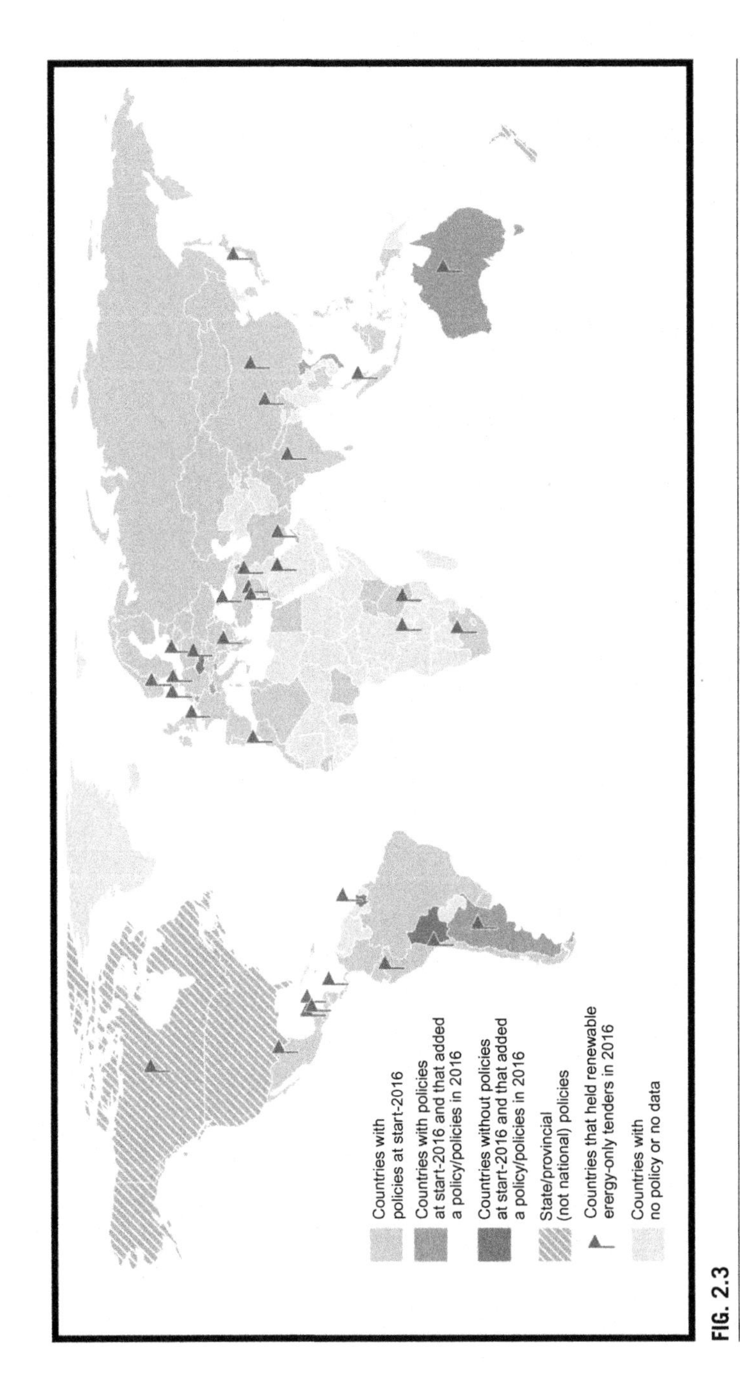

FIG. 2.3

Various countries are focusing on energy production through renewable sources due to the direct correlation of GDP and energy efficiency.

Courtesy: REN21, 2018. Renewables 2018 Global Status Report. Paris: REN21 Secretariat.

heavy metals are also produced during the burning of fossil fuels, giving rise to air pollution, responsible for causing various deleterious effects on health and plant growth [9]. These gases are responsible for producing photochemical smog and acid rain (which leads to the destruction of monuments, crops, and water bodies).

It has been found that intensive coal mining can deprive the land of its vitality, making it barren, with the inability to grow any type of crops. Further, coal mining has resulted in a number of unfortunate deaths in the past, such as the explosion at the Benxihu Colliery (China) in 1942 that killed 1549 miners in a single day. Various reports discuss mass deaths and destruction due to accidents caused by fossil fuels, such as oil spills during transport leading to complete destruction of aquatic ecosystems, the largest spill being Ixtoc I in Mexico that polluted 162 miles of US beaches; fires such as that at an oil depot in Jaipur leading to the deaths of 5 people and injuring over 200; explosions such as the Sinopec pipeline containing natural gas in Shandong Province in China that killed 55 people in 2013, etc. [10, 11].

The renewable sources of energy are plentiful and have no side effects harmful to the environment or human health. They are clean and sustainable, dependable and abundant, and will be cost effective once a suitable infrastructure is well developed. The major sources of renewable energy are solar, wind, biomass, geothermal, hydropower, and tidal energy. It has been estimated that renewable energy sources can generate a pool of new jobs. More and more businesses are investing in the renewable energy sector due to the potential for profit. It is predicted that renewable energy will be a major factor in the future in improving many countries' economies (Table 2.1). In this chapter, the focus has been placed on solar, biomass, and hydropower as major renewable energy sources for various applications. Nanotechnology and enzyme technology have been implemented for generating electricity and for producing fuel in the form of biodiesel, bioethanol, and hydrogen gas (having much higher calorific values than natural gas).

2.3 Overview: clean energy sources in present context

2.3.1 Biodiesel

Fatty ester (FAME), with a long chain of alkyl groups (like methyl, ethyl, or propyl) is prepared using vegetable oil or animal fat followed by the addition of a suitable alcohol. It can be used in its pure form (B100) or can be blended with fuel, such as petroleum diesel at different concentrations for different diesel engine types. For example, B2 refers to 2% biodiesel and 98% petrodiesel, B5 refers to 5% biodiesel and 95% petrodiesel, and B20 refers to 20% biodiesel and 80% petrodiesel, etc. It has been recommended that B20 can be used in diesel equipment with either negligible or minor modifications. In the case of higher blends, engine modification is the primary requisite to avoid maintenance and performance problems. Blends of petroleum diesel and biodiesel are prepared by various methods: mixing in tanks at required concentration prior to delivery at manufacturing point, splash mixing (adding specific percentage of biodiesel and petroleum diesel), and in-line mixing

Table 2.1 Ranking of top countries extracting major source of energy from renewables.

Renewable power	Rank 1	Rank 2	Rank 3	Rank 4	Rank 5
Biopower generation	United States	China	Germany	Brazil	Japan
Geothermal power capacity	United States	Philippines	Indonesia	New Zealand	Mexico
Hydropower capacity	China	Brazil	United States	Canada	Russian Federation
Hydropower generation	China	Brazil	Canada	United States	Russian Federation
Concentrated solar power capacity	Spain	United States	India	South Africa	Morocco
Solar photovoltaic capacity	China	Japan	Germany	United States	Italy
Wind power capacity	China	United States	Germany	India	Spain

Based on data from https://en.wikipedia.org/wiki/100%25_renewable_energy and https://en.wikipedia.org/wiki/List_of_countries_by_electricity_production_from_renewable_sources.

(two components are mixed simultaneously to achieve final blend). Biodiesel has different solvent properties with respect to petrodiesel, having a higher probability of undergoing degradation of natural rubber gaskets and hoses present in vehicles manufactured before 1992. However, this problem has been minimized in recent vehicles through replacement by FKM (fluoroelastomer), which is nonreactive to biodiesel. Further, usage of biodiesel often presents clogging of fuel filters with particulates. It has been recommended to change the fuel filters of engines and heaters periodically when using biodiesel blends. Applications of biodiesel are:

- Replacing fossil fuels: It has been recommended to use biodiesel with B5 in all transport fuel as an essential obligation in the UK and the United States. DaimlerChrysler has increased using biodiesel B20 prepared in the United States at standard level from 2007. The Volkswagen Group has indicated that several of its vehicles are compatible with blends B5 to B100 prepared from rape seed oil with the EN (European Norms) 14,214 standards. Mercedes Benz has allowed B5 to be used in its vehicles and has recommended not to use higher blends due to concerns about shortcomings in production of biodiesel. In 2004, Nova Scotia (in Halifax) began to use city buses running on biodiesel prepared using fish oil. Initially, some mechanical issues have created problems with efficiency, which were rectified later by thorough biodiesel refining. McDonald's (UK) decided in 2007 to produce biodiesel from its waste oil (restaurant by-product) and use it as a fuel to run its fleet [12].
- Railways: Virgin Trains (UK) are known to be the first to run on biodiesel B20. The British Royal Train set a record by running on 100% biodiesel (supplied by Green Fuels Ltd.) on 15 Sept 2007. This train is still running successfully on B100. Similarly, in eastern Washington, a state-owned short-line railroad has been running on biodiesel B25 (from canola) since 2008. Disneyland started their park trains running on B98 (98% biodiesel) in 2007, but this was discontinued in 2008 due to storage issues associated with biodiesel. In January 2009, the park again started running all trains on biodiesel manufactured from used cooking oils as well as soybean oil. The first biodiesel locomotive was added to an all-steam locomotive fleet by Mt. Washington Cog Railway in 2007. In India, the Railways budget announced that B5 (5% biodiesel) would be used in Indian Railways' diesel engines on 8 July 2014 [13].
- Aircraft: Biodiesel was first used in a test flight performed by a Czech jet aircraft. In 2011 United Airlines was the world's first commercial aviation flight to run on biodiesel (derived from algae) as jet fuel. Eco-skies Boeing 737–800 planes have used blend B40 containing 40% biodiesel (derived from algae) and 60% petroleum-derived jet fuel. The Dutch flag carrier KLM (Koninklijke Luchtvaart Maatschappij) began to use biodiesel for all flights departing from Los Angeles International Airport in September 2016 [14].
- Heating oil: Biodiesel has been a very useful heating fuel in domestic and commercial boilers and is often referred to as bioheat fuel (registered trademark of NBB [National Biodiesel Board] and NORA [National Oilheat Research Alliance]

in the United States and Columbia Fuels in Canada). It is generally used as blends of biodiesel and traditional heating oil as per requirements. Generally, blends of higher levels of up to 20% biodiesel are most commonly used by consumers. However, there are some issues with older furnaces while using biodiesel blends due to the presence of rubber parts. Further, these blends can lead to clogging of pipes and fuel filters. It has been suggested that biodiesel blend B20 can be appropriately used as a heating oil, leading to minimization of CO_2 emissions by 1.5 million tons per year (Biodiesel Expo 2006 in the UK). On July 1, 2010, a law was passed under Massachusetts Governor Deval Patrick to use B2 in all home heating. Later, in 2013, this was increased to B5 [15].

- Cleaning oil spills: Oil spills are harmful to aquatic ecosystems. Various technologies have been recommended for cleaning oil spills, but most of them are very costly and quite laborious. Biodiesel has been found to be effective in cleaning oil spills by dissolving them because of the presence of a methyl ester component. Further, it has been found that the dissolving ability of biodiesel is dependent on the type of fatty acids being used for its synthesis. Laboratory experiments have found that higher buoyancy of biodiesel has an important role in effective removal of crude oil. Mixing of oil and biodiesel results in the formation of a thick skim over the water surface, which can be easily removed by various mechanical devices. Further, any leftover mixture (oil and biodiesel) is degraded in a short time due to increased surface area exposure of the mixture [16].
- Biodiesel in generators: The University of California (Riverside) has been using generators based entirely on biodiesel since 2001. Problems like smog, ozone, and sulfur emissions have been solved completely due to complete replacement of generator fuel with biodiesel. These biodiesel generators can be easily used in residential areas around schools, hospitals, and the general public [17].

2.3.2 Bioethanol

Bioethanol is ethyl alcohol used as a fuel, particularly as an additive for gasoline in motor fuel. It is a form of renewable energy produced most commonly from biomass, such as corn, sugar cane, hemp, potato, cassava, etc. However, using agricultural feedstocks for bioethanol production has a negative effect on food prices (substantial increment) due to requirements for large amounts of cultivable land for growing crops suitable for bioethanol production. Therefore, new substitutes have been looked for that are edible and that can be produced in substantial amounts in a small area. Production of ethanol from cellulose (major and universal component in plant cell walls) has been quite lucrative due to substantial amounts of ethanol production with respect to other reported sources. The production of world ethanol for transport fuel tripled from 17×10^9 L to 52×10^9 L in the period 2000–2007. The United States and Brazil are the top producers of ethanol, accounting for 62.2% and 25% of global ethanol production, respectively [18]. Ethanol as fuel is used in a blended state; E10 (10% ethanol) has been recommended for all cars (manufactured after 2000).

Blended ethanol as fuel has been widely used in Brazil, the United States, and Europe. The first car to run entirely on ethanol was a Fiat 147 in 1978 in Brazil. In Brazil, it has been mandatory to blend ethanol with gasoline (most commonly E25, 25% ethanol) for all motor fuels since 1976. Over 14.8 million flex-fuel automobiles and light trucks as well as 1.5 million flex-fuel motorcycles have been using neat ethanol fuel since 2011 in Brazil [19].

2.3.2.1 Applications

- Automobile fuel: Comparative analysis on energy per unit volume of ethanol and gasoline has revealed that the former has ~34% less energy. However, ethanol has a higher octane rating, promoting better engine efficiency due to an increased compression ratio. Lower ethanol blends have increased the efficiency of motor fuel, but higher blends (>E25) generally result in lower mileage and require frequent refueling, depending on vehicle type. Based on EPA (Environmental Protection Agency) testing, it was found that usage of ethanol blend E25 resulted in lowering of average fuel economy by ~30% with respect to unleaded gasoline in all 2006 models. However, E85 has been found to be a high-performance fuel with an octane rating of 94–96. A thermal efficiency equivalent to that of diesel can easily be achieved by using fuels ranging from neat ethanol to blends of E50 having a compression ratio of 19.5 [20]. Since 1999 there has been a tremendous increase in the number of vehicles manufactured worldwide that can run on ethanol, even up to 100% ethanol without modification, due to the absence of alcohol sensors. In Sweden, ethanol engines have been designed since 1989 based on the diesel principle and used primarily in city buses, distribution trucks, and waste collectors. In Scandinavia, engines have been modified with respect to compression ratio, leading to usage of mixed ethanol blends of E95 containing 93.6% ethanol, 3.6% ignition improver, and 2.8% denaturants. Possibly the most important innovation has been the development of dual-fuel direct-injection involving efficient mixing of ethanol and gasoline in accurate ratios using turbocharged, high compression ratio, and small displacement engines. Here, ethanol is taken into a smaller tank and injected directly into the cylinders along with gasoline, leading to higher octane ratings up to 130 with reduction of CO_2 emission by 30%. Further, there is an improved consumer cost payback time, by about 25%, with resolution of problems like water absorption, cold weather starting, etc.
- Cold start during the winter: Starting the engine in low temperatures has been a frequent problem associated with using ethanol-based fuel due to lowering of vapor pressure below 45 kPa (especially at temperatures lower than 11 °C). It has been recommended in the United States and Europe to use ethanol blends lower than E85 in fuel vehicles. However, these ethanol blends are further lowered (E75) when temperatures fall below −12 °C during winter. Further, it has been recommended to install an engine heater system while using ethanol blends during the winter months. In Brazil, flex-fuel vehicles have been commonly used that operate with ethanol E100 (hydrous ethanol with up to 4% water).

Here, lowering of vapor pressure has been much faster with respect to blend E85. Therefore, in these flex vehicles a small secondary gasoline reservoir is kept near the engine. During winters, when temperatures fall below 15 °C, pure gasoline is injected and this helps with cold start problems [21].

- Fuel cells: Since 2016 several car manufacturers, like Nissan Hyundai Tucson FCEV, Toyota Mirai, and Honda FCX, have developed fuel-cell vehicles that use liquid ethanol fuel as a source to generate hydrogen within the vehicle itself. This technology uses heat to convert ethanol into hydrogen (solid oxide fuel cell, SOFC) and generates electricity to supply power to electric motors driving the wheels. Nissan uses an ethanol blend of E55 for their fuel cell and even announced that they would commercialize this technology by 2020 [22].

2.3.3 Solar cell/photovoltaics

Solar cells/photovoltaics (PVs) have been very successful in generating electricity through conversion of light into electricity via semiconductors that exhibit the photovoltaic effect. They have been used since 1990, but solar panels began to be mass-produced when German environmentalists and the Eurosolar organization received government funding for a 10,000-panel roof program in 2000 [23]. Solar panels are used for absorbing sunlight and converting it into electricity followed by conversion of the electric current from DC to AC via solar inverters. Solar panels (containing a number of solar cells) can be easily mounted on the ground, rooftops, or walls aligned towards direct sunlight, with capacities from a few to several tens of kilowatts (kWs) to hundreds of megawatts (MWs) in large utility-scale power stations. Installations of solar panels can operate for over 100 years with minimal maintenance in addition to affordable operating costs with respect to existing power technologies. Solar PVs do not produce any type of pollution or greenhouse gas emissions. Further, recycling technologies have been developed to maximize the use of end wastes generated from usage of solar PVs. They have very simple scalability depending on power requirements. Silicon is used in PVs and is available in abundance on earth, which further promotes using solar PVs for generation of electricity. However, there is comparatively smaller research investment on the development and improvement of solar cells with respect to research on fossil and nuclear energy sources. Present experimentation on solar cells has achieved over 40% efficiencies and this is still rising with the advent of more and more technologies leading to further lowering of production costs of solar cells.

This chapter discusses how nanotechnology and enzyme technology have been used for enhanced efficiency and lowering of cost in the production of solar cells. It is assured that solar energy will become a primary energy source in the coming years. Research has also focused on how to reduce the transmission/distribution losses in grid-connected solar electricity. Solar PVs have been adopted by over 100 countries, and the governments of many countries are giving lucrative incentives for solar PV installations, to maximize usage of solar energy. It has been estimated that solar PVs installed worldwide will provide over 500 GW of electricity (~5% of global electricity demand) by the end of 2020. China, Japan, and the United States are the fastest-growing markets for solar PVs, with Germany being the world's largest producer of solar PVs [24].

2.3.3.1 Applications

- Rooftop and building integrated systems: PV arrays are being mounted on building rooftops or nearby on the ground, or sometimes integrated into the building. Based on the position of the PV installation, they are categorized as rooftop PV systems, mounted on top of the existing roof structure or on the walls, or building-integrated PVs (BIPVs), incorporated into the roof/walls of the buildings. Sometimes PVs are installed inside the roof tiles to have more open space. Rooftop PV systems are the more useful: besides producing electricity, they provide passive cooling on buildings during the day and accumulate heat for the night. Further, rooftop systems generate more electricity, ranging from 10kW (residential) to several hundred kW (commercial). Advanced PV technologies have been developed, viz. CPVs (concentrator photovoltaics) in which lenses and curved mirrors are added for focusing sunlight onto small, efficient multijunction solar cells; and PVTs (PV thermal hybrid solar collectors), which convert solar radiation into thermal as well as electrical energy, thus providing higher energy efficiency [25].
- Power stations: The Solar Star PV power station produced 579 MW (MW_{AC}) in 2015 and became the world's largest photovoltaic power station at that time, followed by the Desert Sunlight Solar Farm and the Topaz Solar Farm (both with a capacity of 550 MW_{AC}), all constructed by US companies. All three power stations are located in the California desert. These power stations produce no emissions and have no fuel costs during their operation [26]. Larger solar power stations have come online since 2015 and additional larger plants are proposed at various sites around the world.
- Rural electrification: Rural areas of many developing countries are now using grid power generated using PVs. In various areas of India, rural lighting programs using PV technology have provided solar-powered LED lighting, supplanting kerosene lamps. Before 1995, solar rural electrification projects were economically unfavorable and lacked the necessary technical support, with many complexities in technology transfer for various administrative reasons. The most common applications of PV technology in remote areas include: solar lamps, emergency telephones, water pumps, temporary traffic signs, parking meters, trash compactors, charging stations, and remote guard posts and signals.

The Floatovoltaic system was developed in May 2008 by the Far Niente Winery (Oakville, CA), consisting of 994 PV solar panels installed onto 130 pontoons floating on the winery's irrigation pond, with generation of ~477kW of peak output. This system is interesting because it does not consume any land area for the PV installation; thus land area can be used for agriculture or other purposes. The Floatovoltaic system has an additional benefit of lowering the temperature of the solar panels being used for electricity generation and making them more efficient. The floating panels also inhibit algal growth on the water surface due to the water evaporation process [27].

- Transport: Presently, PV is used only rarely in transport applications, but the idea of solar-powered air conditioners in automobiles has been recently put forward. Solar-powered cars, boats, and airplanes have been found to be efficient and long-lasting. The Solar Impulse 2 (Swiss solar aircraft) achieved the longest nonstop solo flight in 2015.
- Telecommunication and signaling: Solar PVs are used in local telephone exchanges, radio and TV broadcasting, microwave, and other forms of electronic communication links. In mountainous regions, radio and TV signals are generally blocked or reflected due to undulating terrain. In these locations, low-power transmitters (LPTs) based on solar PVs are helpful in receiving and retransmitting signals.
- Spacecraft applications: Spacecraft are one of the earliest applications of PVs, beginning with silicone-based solar cells used on the Vanguard 1 satellite (launched by the United States in 1958). Basically, solar cells are used to power sensors, communications, and active heating and cooling. Sometimes solar cells are used in spacecraft propulsion. Solar cells have been an essential component in important missions such as the MESSENGER probe to Mercury, the Junoprobe to Jupiter, etc. Continuous innovations include increased amount of power generated, cost reductions, and higher efficiency, as well as discovery of more effective semiconductors like gallium arsenide (GaAs) [28].

2.3.4 Hydrogen

Hydrogen is an ideal replacement for fossil fuels in various applications. It is a potential candidate to replace conventional power generators like combustion engines, large batteries in cars, buses, forklift trucks, submarines, power plants, etc. Further, efforts are being made towards using hydrogen as an alternative fuel for empowering mobility and transportation needs. Hydrogen is three times more powerful and efficient than gasoline and other fossil fuels, due to production of more energy per pound of fuel. Further, it can generate electricity between 33% to 35% efficiency in power plants, and 65% in the case of fuel cells. It is also used by industry for refining petroleum, metal treatment, production of fertilizers, and in food processing. Hydrogen is not available on earth in a free state but is present in bound states, viz. bonded to oxygen in water, and to carbon in live, dead, and/or fossilized biomass. Common methods like electrolysis are used to produce hydrogen by splitting water from stream, hydrocarbons, or through acidic reactions of metal hydrides, etc. Even though hydrogen is an excellent renewable source, its production is dependent on fossil fuels [29] (48% from natural gas, 30% from oil, 18% from coal) with 4% from electrolysis [29]. Further, this is also the reason for the costliness of hydrogen production. Various alternative methods are being explored to produce hydrogen in substantial amounts without using fossil fuels. This chapter discusses the use of nanotechnology and enzyme technology for economical hydrogen production.

Hydrogen can be stored either in a liquid or solid state with a requirement of high-pressure tanks with pressure ranging from 350 to 700 bars. In the case of hydrogen storage in a liquid state, cryogenic temperatures are needed, i.e., −252.8 °C at 1 atm. Thus, hydrogen is more demanding and laborious to produce and store than other energy sources. However, it is known to be the cleanest source of energy due to nonproduction of harmful by-products. It produces water on burning in addition to release of energy for various applications. Thus, it is an ecofriendly fuel with negligible toxicity.

2.3.4.1 Applications

- Vehicles: Hydrogen can be used as a transportation fuel due to its ability to power fuel cells in zero-emission electric vehicles. Over 500 types of hydrogen-fueled vehicles are being used in the United States, including cars, buses, and forklift trucks, with the help of numerous industrial partnerships working for the reduction of carbon emissions. Cities like San Francisco and Berlin have developed a fueling infrastructure to promote more hydrogen-based vehicles.
- Submarines: Hydrogen-powered submarines have an exceptional ability to be under water for much longer times than conventional submarines. Hydrogen is even being used in ships, which solves various problems associated with heavy oils. Hamburg, Germany supplies hydrogen through various fueling stations to boats and ships.
- Rocket fuel: NASA (The US National Aeronautics and Space Administration) is the largest user of hydrogen as rocket fuel. It has begun using liquid hydrogen for fuel and is the first to use hydrogen-based fuel cells to power the electrical systems on spacecraft.
- Used in backup and off-grid power: Hydrogen is used in production of electricity through the combination of hydrogen and oxygen atoms. Hydrogen-based fuel cells produce electricity that can be used in small-scale applications (powering laptop computers, cell phones, and various military applications) to large-scale applications (providing electricity for emergency power in buildings and remote areas without any power lines). The Hymera generator, the first hydrogen-based portable power device, developed in Germany, is used for producing electricity on a large scale.

2.4 Nanotechnology in production of clean energy

A variety of nanomaterials are being used for extracting energy in an efficient way from different renewable sources (see Table 2.2).

2.4.1 Biodiesel production using nanoparticles

Biodiesel, also called FAME (fatty acid methyl ester), can be produced from fatty acids of various waste products like vegetable oil and animal fat by using the appropriate catalysts. In this section, the production of biodiesel using nanoparticles

Table 2.2 Nanomaterials used in extracting energy from various renewable sources efficiently.

Biodiesel	Bioethanol	Photovoltaics	Hydrogen
KF/γ-Al_2O_3 NPs	Co-nanocatalyst	SiO_2@TiO_2 colloidal NCs	Pt NPs & TiO_2 NTs
ZrO_2/Al_2O_3 NPs	$CoFe_2O_4$@ SiO_2-CH_3 NPs	SnO_2/TiO_2 hybrid NS	Pd(shell)-Au(core) NPs
Al-Sr NCs		TiO_2 NRs	N-doped TiO_2 nanofibers
Mg-Al hydrotalcite (HT-Ca) NPs		Flower-shaped TiO_2	TiO_2 NTs
Al/Y_2O_3 NPs		MWCNTs	Fe_2O_3/TiO_2 NC
CaO and MgO NPs		rGO/mTiO_2 NC	CdS NPs
MnO_2 NRs		TiO_2 with C_{60} & porphyrin NPs	P25 TiO_2 (85% anatase, 15% rutile)
Li–CaO NPs		SiO_2 NPs	TiO_2 NPs
Li–MgO NPs		ZnS/Si NPs	Graphene-$Zn_xCd_{1-x}S$ NC
Ca/Fe_3O_4.SiO_2 NCs		ZnO/SiO_2 NPs	ZnS NPs
Au/CuO NCs		$Cu_2ZnSn(S, Se)_4$ NPs	ZnS/Au NPs
Fe_3O_4-Na_2SiO_3 NPs		n-type ZnO Nanowires	AgInZnS-reduced graphene NC
Fe_3O_4@silica magnetic NPs		c-Si/Ag NPs	platinum-palladium (Pt–Pd) NP alloy
$Cs_xH_{3-x}PW_{12}O_{40}$/ Fe-SiO_2 NCs		Si & Ge NPs	Nanoferrite
[Ti(SO_4)O] NCs		Au@SiO_2 NPs	Flower-shaped CdS NPs
TiO_2/PrSO_3H NCs		FeS_2 NCs	Mo_2C NPs
TiO_2-ZnO NCs		Au & [P_3HT]: [PCBM] NPs	MoO_{3-x} NPs
ZnO-Na NCs		PbS_xSe_{1-x} NCs	Nitrogen-doped, cobalt-encased CNTs
Cu-ZnO NCs		$Y_2O_3:Er^{3+}/Yb^{3+}$ NPs	Ru NPs
Fe-Mn- MoO_3/ZrO_2 NCs			$CuGa_2In_3S_8$ NCs
ZrO_2/CdO NCs			Nanosilicon
SO_4^{2-}-ZrO_2 NPs			
ZSM-5 aluminosilicate zeolite NCs			
La_2O_3 NPs			

for greater efficiency and stability is discussed. Further, nanotechnology has also conferred cleaner production of biodiesel without generation of toxic by-products.

Biodiesel has been synthesized by using KF-doped nanoparticles of γ-Al_2O_3 as a catalyst for the transesterification of vegetable oil with methanol. The effective parameters like catalyst concentration, KF to nano-γ-Al_2O_3 ratio, calcination temperature, molar ratio of methanol/oil, temperature of the transesterification reaction, and time have been optimized suitably to achieve a high conversion of vegetable oil to biodiesel [30]. In another report, KF-doped nanoparticles of γ-Al_2O_3 were used for the transesterification of a mixture of canola oil in the presence of dimethyl carbonate and methanol and produced biodiesel with a yield of 98.8% [31]. Heterogenous ZrO_2/Al_2O_3 nanoparticles with active catalysis were prepared to convert oleic acid and methanol into fatty acid esters (biodiesel) under high voltage conditions in a low temperature and atmospheric pressure process [32]. The Al–Sr nanocatalysts were obtained after calcination at 600 °C for 5 h by a coprecipitation method. They were utilized for the transesterification of sunflower oil with a biodiesel yield of 96.8% [33]. Activated Mg–Al hydrotalcite (HT-Ca) nanoparticles with a size of <45 nm have been synthesized, having both acidic and basic properties due to formation of $Mg_4Al_2(OH)_{14}{\cdot}3H_2O$, $Mg_2Al(OH)_7$ and AlO(OH) nanocrystals. Five percent of HT-Ca nanocrystals are used to esterify oil with high acid value (AV); viz. soybean oil having an AV of 12.1 with molar ratio of methanol/oil (30/1) at 160 °C for 4 h resulted in a yield of 92.9% [33]. Mesoporous, amorphous nano-alumina powder containing yttrium oxide synthesized using the sol gel method was found to produce a very high percentage biodiesel-to-oil ratio via the transesterification reaction [34].

Nano-CaO and -MgO have been synthesized using the sol gel method and were used in a transesterification reaction to produce biodiesel as well as glycerin [35]. Nanorods of MnO_2 were used in transesterification of palm oil when the methanol-to-oil molar ratio was 9:1 at 60 °C [36]. A novel heterogenous nanocatalyst, CaO, has been synthesized using eggshells (waste biomass) with a size of 75 nm; it was used for transesterification of dry biomass into biodiesel, with a yield of 86.41% [37]. In another report, a nanocatalyst (dia. 27–16 nm) with regular spheroidal structure was prepared using eggshells of *Gallus domesticus* [38]. The catalytic activity of the prepared nanocatalyst resulted in production of biodiesel (FAME yield: 97%) via glycerolysis and transesterification with methanol of crude *Jatropha curcas* oil. Li-impregnated CaO nanoparticles were synthesized and used as a catalyst for transesterification of Karanja and Jatropha oils containing 3.4 and 8.3 wt% of free fatty acids, respectively [39]. The impregnation of Li has also been tried with MgO nanoparticles and used as a catalyst for transesterification of mutton fat into biodiesel. It has been found that impregnation with potassium (3.5 wt%) onto MgO nanoparticles led to the best response and complete transesterification, even in the presence of 10.26 wt% moisture and 4.35 wt% free fatty acid content [40]. Magnetic $Ca/Fe_3O_4.SiO_2$ nanocatalysts (diameter ~50 nm) prepared by sol gel and incipient wetness impregnation have been used for the production of biodiesel with a yield of 97% [41]. A novel strategy has been adopted to synthesize 1D Au/CuO nanocatalysts in which there is uniform distribution of Ca^{2+} and Au^{3+} ions through introduction of a coordination polymer via chemical bonding with bifunctional organic linkers. Prepared Au/CaO nanocatalysts were in the form of ribbons, exhibiting excellent catalytic performance in the transesterification reaction due to their large surface area. It has been reported that these nanoribbons are recyclable with excellent efficiency in biodiesel production with respect to other Ca^{2+} based nanocatalysts [42].

Biodiesel has been prepared by transesterification of cottonseed oil using magnetic Fe_3O_4 nanoparticles loaded with Na_2SiO_3, leading to an excellent yield of 99.6%. Here the transesterification conditions so optimized were: methanol/oil molar ratio of 7:1, 5% of catalyst dosage, reaction temperature of 60 °C, reaction time of 100 min, with stirring speed of 400 rpm [43].

Core-shell Fe_3O_4@silica magnetic nanoparticles have been synthesized and functionalized with triazabicyclodecene (TBD). These nanoparticles were used as catalysts for producing biodiesel from microalgae oil sources (dried algae, algae oil, and algae concentrate). TBD-functionalized Fe_3O_4@silica nanoparticles have effectively converted algae oil into biodiesel with a yield of 97.1%. This presents an excellent example of utilizing algal biomass and converting it into biofuel in the form of biodiesel [44]. A nanocatalyst, $Cs_xH_{3-x}PW_{12}O_{40}/Fe\text{-}SiO_2$, has been prepared by using the sol gel method in the presence of experimental conditions as follows: reaction temperature 323–333 K, methanol/oil molar ratio = 12/1, and reaction time 0–240 min. The basis of preparing this nanocatalyst and using it for transesterification for biodiesel production is that heteropolyacid $H_3PW_{12}O_{40}$ has been known for its efficient ability to produce biodiesel in either homogeneous or heterogeneous catalytic conditions. It has also been indicated that $Cs_xH_{3-x}PW_{12}O_{40}/Fe\text{-}SiO_2$ (magnetic nanocatalyst) can be easily recycled more than five times without any loss of catalytic activity [45].

Titanium-based nanocatalysts have been used for biodiesel production in various forms. A bidentate sulfate coordinated solid acid nanocatalyst [$Ti(SO_4)O$] has been used for the esterification/transesterification of free fatty acids obtained from cooking oil for the production of biodiesel [46]. A FAME yield of 97.1% was achieved using this catalyst for the esterification of used cooking oil with a recyclability of 8, as indicated by the presence of SO_4^{2-} species in the XPS (x-ray photoelectron spectroscopy) spectra. The mesoporous $TiO_2/PrSO_3H$ solid acid nanocatalyst has better recyclability with respect to [$Ti(SO_4)O$] due to the presence of a propyl sulfonic acid group. It has a FAME yield of 98.3% using the same cooking oil as used by [$Ti(SO_4)O$] [47].

Biodiesel has been produced by transesterification of palm oil by mixing with oxides of TiO_2-ZnO nanocatalysts. Here, stable and active heterogeneous mixtures of TiO_2-ZnO and ZnO nanocatalysts have been prepared for the development of a simpler transesterification process. These nanocatalysts have been recommended for biodiesel production at industrial scale due to their excellent catalytic performance with reduced temperature and time. A biodiesel yield of 98% was reported at optimum reaction parameters with 200 mg of nanocatalyst [48].

The nanocatalyst ZnO impregnated with sodium (5 wt%) has been used for transesterification of completely virgin cottonseed oil with methanol [49]. The catalytic activity of the ZnO nanocatalyst depends on: amount of impregnated sodium, calcination temperature, methanol-to-oil molar ratio, reaction temperature, and the amount of free fatty acid content in the feedstock. This nanocatalyst has also been used for the transesterification of various other feedstocks like virgin cottonseed oil, used cottonseed oil, mutton fat, Karanja oil, and Jatropha oil, having free fatty acid contents in the range of 0.1–8.5 wt%. CZO (Cu-doped ZnO) nanocatalysts have been prepared to transesterify neem oil. The reaction conditions chosen to obtain the maximum biodiesel yield of 97.18% were: reaction time (60 min), reaction temperature (55 °C), CZO nanocatalyst (10% *w*/w) and oil-methanol ratio (1:1; v:v). The nanocatalyst is recyclable up to six cycles without any loss in catalytic performance [50]. Another report on a CZO nanocatalyst was based on production of biodiesel using cooking oil waste. Here nanocatalysts were in the form of nanorods with a size of 80 nm. Reaction conditions were: nanocatalyst concentration (12% w/w), oil-methanol ratio (1:8; v:v), reaction temperature (55 °C), and reaction time (50 min); these resulted in a biodiesel yield of 97.71% [51].

The solid acid nanocatalysts FMMZ (Fe–Mn– MoO_3/ZrO_2) were prepared and used for the production of biodiesel from waste cooking oil. Here the focus was placed on the influence of ferric and manganese dopants on the catalytic efficiency of synthesized nanocatalysts. The nanocatalyst was used to convert waste cooking oil into biodiesel, with a yield of 95.6% corresponding to a reaction time of 5 h at 200 °C with a stirring speed of 600 rpm, with oil to alcohol ratio of 1:25 in the presence of 4% w/w catalyst [52]. Zirconia-supported CdO nanocatalysts have also been used for transesterification of soybean oil with methanol into biodiesel with reaction kinetics following a pseudo–first order equation with activation energy (E_a) of 41.18 kJ/mol. A biodiesel yield of 97% was achieved using 7% catalyst loading and 1:40 M ratio of oil to methanol at 135 °C [53]. The nanoparticles of SO_4^{2-}-ZrO_2 have been used as a solid acid catalyst for conversion of free fatty acids into biofuels (biodiesel) [54]. Impregnation of the SO_4^{2-} group has been useful for improving catalytic performance of the nanoparticles as well as in monodispersion. These nanoparticles have demonstrated effective conversion of palmitic acid into methyl palmitate (biodiesel). Further, addition of surfactants like poly (N-vinylpyrrolidone) was found to improve synthesis of biodiesel via an electrophilic substitution reaction of indole with aldehydes [55].

The nanocrystalline ZSM-5 (zeolite socony mobil with framework type 5) is an aluminosilicate zeolite belonging to the pentasil family of zeolites with the chemical formula $Na_nAl_nSi_{96-n}O$); it has exhibited excellent catalytic properties, such as moderate acidity and stacking order of mesoporosity, responsible for the production of gasoline with a high selectivity (77.4%) and 95 research octane number [56]. This zeolite-based nanocatalyst has been used to produce green gasoline from bio-oil derived from the *Jatropha curcas* plant. A novel method has been adopted in synthesizing nano-La_2O_3 using a sonochemical and hydrothermal method. It is then used in the transesterification reaction to produce biodiesel [57]. It has been found that nano-La_2O_3 synthesized by the sonochemical method has showed relatively higher catalytic activity, which may be ascribed to its high base strength, large base amount, small particle size, and large BET (Brunauer Emmett Teller) surface areas. A hierarchically integrated CNT-based nanocatalyst has been utilized for the production of biodiesel from bio-oils as well as for other more valuable chemicals, like glycerol, simultaneously [58].

2.4.2 Bioethanol production by nanoparticles

Spent tea (*Camellia sinensis*) has been converted into bioethanol by utilizing three steps consecutively, with a conversion rate of 57.49% [59]. The whole process involves gasification of spent tea using a Co nanocatalyst at 300 °C, followed by transesterification of liquid extracts, and subsequently *Aspergillus niger* is grown on leftover spent tea. Thus, an effective way of utilizing waste tea for the production of bioenergy as an alternate energy source has been demonstrated. Methyl-functionalized silica and methyl-functionalized cobalt ferrite–silica ($CoFe_2O_4$@SiO_2-CH_3) nanoparticles have been used to produce syngas (H_2 and CO), ethanol, and acetic acid. Here, nanoparticles have been used as catalysts during fermentation of biomass by *Clostridium ljungdahlii*. $CoFe_2O_4$@SiO_2-CH_3 nanoparticles have a recyclability of 5, with effective mass transfer enhancement during fermentation [60].

2.4.3 Energy production by PV cell using nanoparticles

There is an increasing demand to find more cost-effective materials with the photovoltaic effect, to lower the production costs of PVs. Several nanoparticles have been explored in recent years which either have the photovoltaic effect or can be blended with existing materials (i.e., semiconductors) to enhance their efficiency by several hundredfold. It has been found that both organic and inorganic nanomaterials can be used in varied shapes, ranging from spherical, rod-like, or branched. The following section describes some applications of nanoparticles in PVs.

Most of the reports in the past few decades are based on solid-state photovoltaic cells working on the principle of plasmon-induced charge separation (PICS). However, these photovoltaic cells have poor conversion efficiency due to a variety of factors. Simpler plasmonic photovoltaic cells containing interconnected Au, Ag, and Cu half-shell arrays with SiO_2@TiO_2 colloidal nanocrystals deposited on their surface have improved absorption of plasmonic light and current collection due to their excellent PICS (peripheral interface controllers) efficiency. Furthermore, it has been found that cells with Ag half-shell arrays have better photovoltaic performance than cells with Au and Cu half-shell arrays. This is because of the photogeneration of large amounts of energetic electrons, resulting in high injection efficiency with suppressed charge recombination probability [61]. Solar cells have been produced containing SnO_2 nanosheet-structured films prepared on a fluorine-doped TiO_2 (FTO) substrate with a ZnO nanosheet as template. Doping helps in filling nanovoids present at suitable intervals on SnO_2 nanosheets. The prepared hybrid films of SnO_2/TiO_2 are further cosensitized with CdS and CdSe quantum dots. It has been found that sensitized solar cells assembled with SnO_2/TiO_2 hybrid films have excellent photoelectricity conversion efficiency with respect to pure TiO_2 films [62]. Comparative studies on photovoltaic properties of fluorine-coated TiO_2 electrodes prepared with nanoparticles and nanorods have revealed various facts. TiO_2 nanorods–based dye-sensitized solar cells (DSSCs) have exhibited better conversion efficiency and current density with increments of 35% and 38%, respectively. This is so because of faster electron transport and reduced charge recombination [63]. A novel hierarchical flower-shaped TiO_2 has been synthesized containing anatase TiO_2 nanotubes collected on a Ti foil substrate. The photovoltaic performance of these hierarchical TiO_2 flowers is 7.2%, which is much higher than that of TiO_2 nanoparticles (6.63%), due to superior light-scattering ability and fast electron transport. Further, when these novel hierarchical TiO_2 flowers are electrodeposited with poly(3,4-ethylenedioxythiophene) (PEDOT), there is an increase in power conversion efficiency by 6.26%, with a short-circuit current density of 11.96 mA/cm^2 and an open-circuit voltage of 761 mV [64]. The influence of multiwalled carbon nanotubes (MWCNTs) and TiO_2 nanoparticles on photovoltaic performance has been studied using a Grätzel cell with an I^{3-}/I^- redox couple containing electrolyte. It was found that there is increased acceleration of electron transfer in the case of MWCNTs, resulting in enhanced PV performance of the cell with an increase in power conversion efficiency by 60% [65]. Solar cells containing reduced graphene oxide (rGO)/mesoporous (mp)-TiO_2 nanocomposite have been prepared. These solar cells are found to have improved electron transport properties due to reduced interfacial resistance. Here an optimal amount of rGO is added to the TiO_2 nanoparticle–based electron transport layer, being suitably optimized to obtain excellent film resistivity, electron diffusion, recombination time, and photovoltaic performance. The nanocomposite (rGO/mp-TiO_2) film leads to increased photon conversion efficiency, by 18% as compared to TiO_2 nanoparticle–based perovskite solar cells [66]. In another report, TiO_2 nanoparticles were modified by introducing C_{60} and porphyrin units on their surface, and then electrophoretically deposited onto an ITO/SnO_2 electrode. This modification led to maximum incident photon-to-current efficiency of 88% at 410 nm (the highest value reported so far among molecule-based photovoltaic cells). The presence of C_{60} and porphyrin units on TiO_2 nanoparticles enhances the chemical attachment between individual nanoparticles, which improves the effective electron transfer between the photoactive units and the electrodes [67].

Flexible solar cells have been fabricated using a crystalline p-Si/n-ZnO heterostructure. The presence of spherical SiO_2 nanoparticles produces more efficient light entrapment, while ZnO helps in large band bending, so is responsible for improved open circuit voltage. Both factors collectively led to enhancement of base efficiency by ~52%. These solar cells are very cost effective and flexible, and they find extensive applications in next-generation photovoltaics and roll-to-roll electronics [68]. A highly efficient hybrid photovoltaic cell has been prepared based on ZnS nanoparticles/Si nanotips p-n active layer. The photovoltaic cell has improved performance due to the enhanced external quantum efficiency, photoluminescence excitation spectrum, photoluminescence, and reflectance. It can work not only in the ultraviolet region but also can be modulated to work in any given spectral regime [69].

DSSCs have been designed using a hybrid composite of ZnO nanoparticles (ZnO NPs) and silica nanospheres (SiO_2 NSs). The addition of SiO_2 NSs has boosted light-harvesting efficiency and enhanced the photovoltaic performance of the DSSCs. A total of ≈22% enhancement has been observed in the overall power conversion efficiency by a hybrid composite containing DSSCs, which is much higher than solar cells based on ZnO NPs. It has been further reported that SiO_2 NSs are also responsible for serving as a partial energy barrier layer to retard interfacial recombination of photo-generated electrons at working electrodes as well as at the electrolyte interface, which are additional factors in enhancement of device efficiency [70].

High-efficiency thin-film solar cells have been prepared based on binary and ternary chalcogenide nanoparticles composed of $Cu_2ZnSn(S, Se)_4$ (CZTSSe). Laboratory-scale photovoltaic cells have shown total area efficiency of 8.5% of total area with active area efficiency of 9.6% without any antireflective coatings. The presence of a bilayer microstructure has improved thermal processing [71]. Novel solar cells have been prepared containing vertically oriented n-type ZnO nanowires surrounded by a film made of p-type cuprous nanoparticles. This is an inexpensive and environmentally benign solar cell with easily adaptability for production of large-scale photovoltaic devices. The improved performance of this solar cell has been attributed to the presence of an intermediate oxide insulating layer present between the nanowires and nanoparticles [72].

The focus is always on manufacturing solar cells that are cost effective. An easily manufacturable and inexpensive solar cell has been synthesized containing a transparent conductive electrode made of crystalline silicon (c-Si) with a silver nanoparticle network in between the pyramids of the textured solar cell surface. Prepared solar cells have a power conversion efficiency of 14% less than the conventionally processed c-Si control solar cell. However, the significantly lower manufacturing cost has made these cells commercially viable [73]. There have been many efforts to enhance the efficiencies of solar cell significantly in various ways. In one report, an increased number of bandgaps in multijunction solar cells (helps in effective charge separation) have been used for efficiency enhancement. Nanostructured semiconductors in two forms, silicon (Si) and germanium (Ge) nanoparticles embedded in wide gap materials and mixed silicon-germanium nanowires, were tried and compared for efficiency. It was found that in the former case, there is a clear interface between Si and Ge, which reduces the quantum confinement effect with a natural geometrical separation between the electron and hole. Thus, the former combination has much higher efficiency than the latter [74]. Meso-superstructured organometal halide perovskite solar cells containing core-shell Au@SiO_2 NPs have shown improved photocurrent and efficiency (~11.4%). Here, the enhanced photocurrent is attributed to reduced exciton binding energy with enhanced light absorption [75].

Iron pyrite (FeS_2) nanocrystals (NCs) have been found to have a large absorption coefficient, low cost, and great abundance. Thus, they are potential candidates to be used in synthesis of photovoltaic cells. It has been reported that p-type colloidal FeS_2 NCs have localized surface plasmon resonances (LSPRs) responsible for plasmonic photoelectron conversion. A well-defined percolation network has been prepared that contains heterojunction nanostructures made of FeS_2 NCs (80 nm) and CdS quantum dots (QDs, size 4 nm) to enhance power conversion efficiency under AM 1.5 solar illumination with open circuit voltage of 0.79 V [76]. A single-layer graphene film has been fabricated with Au nanoparticles and poly(3,4-ethylenedioxythiophene) [P_3HT]: poly(styrene sulfonic acid) [PCBM] and used in preparing solar cells. A fabricated single layer of graphene and indium TiO_2 (ITO) was used at the top and bottom of the conducting electrodes, respectively. It was reported that a maximum efficiency of 2.7% was achieved by using a graphene electrode with surface area of 20 mm^2 under an AM1.5 solar simulator. Further, power conversion efficiency has been found to decrease from 3% to 2.3% when the active area of graphene was increased from 6 to 50 mm^2 due to increased series resistance and decreased edge effects [77].

Photovoltaic cells based on highly confined nanocrystals of ternary compound (PbS_xSe_{1-x}) were found to be more efficient than those based on pure PbS or PbSe [78]. The efficiency of DSSCs was found to be improved by using Er^{3+}/Yb^{3+} co-doped with Y_2O_3 and forming $Y_2O_3:Er^{3+}/Yb^{3+}$ phosphor nanoparticles with sunlight harvesting in the near-infrared (NIR) region, particularly at 980 nm. These nanoparticles have improved power conversion efficiency due to their effective light-harvesting efficiencies. DSSCs based on TiO_2 films as photoelectrode and $Y_2O_3:Er^{3+}/Yb^{3+}$ nanoparticles (light-harvesting center) have a power conversion efficiency (PCE) of 6.68% with a photocurrent of $J_{SC}=13.68\ mA/cm^2$ [79].

2.4.4 Production of hydrogen by nanoparticles

A bioinspired nanocomposite has been prepared containing Pt nanoparticles and fabricated TiO_2 nanotubes with natural cellulose. The nanocomposite has a 3D hierarchical structure with ultrafine metallic Pt nanoparticles (size: 2 nm) that have been uniformly immobilized onto the surface of TiO_2 nanotubes. This can produce 16.44 mmoL $h^{-1} g^{-1}$ of H_2 by water splitting through a photocatalytic process when 1.06 wt% of Pt content is present in the nanocomposite. It has been reported that the amount of H_2 produced is drastically reduced on adding an excessive amount of Pt (i.e., >1.06 wt%) [80]. The novel structure of nanoparticles containing Pd(shell)–Au(core) nanoparticles immobilized on TiO_2 has been synthesized and used for H_2 evolution through photocatalysis. These novel nanoparticles have exhibited high quantum efficiency towards H_2 production by using a wide range of alcohols, thus presenting an additional advantage of using chemical by-products from the birefinery industry. Further, their additional feature of reusability makes them good candidates in various industrial processes. Enhanced performance in H_2 production has been attributed to the presence of unoccupied d-orbital states, which are used for photoexcited electrons in redox reactions in the presence of visible light [81]. N-doped TiO_2 nanofibers have been synthesized with high photocatalytic efficiency in generating hydrogen by using ethanol-water mixtures under UV-A and UV-B irradiation. It has been reported that 100 mg of N-doped TiO_2 nanofiber in the presence of 1 L of water-ethanol mixture has produced 700 μmoL h^{-1} and 2250 μmoL h^{-1} of

H_2 in the presence of UV-A and UV-B, with photo energy conversion of ~3.6 and ~12.3%, respectively [82]. Nanotube arrays of TiO_2 (TNAs) have been prepared and sensitized with silver sulfide (Ag_2S) nanoparticles (NPs) followed by in situ sulfurization to form Ag/TNAs. The synthesized nanocomposites have photoconversion efficiency of 1.21% with 1.13 $mLcm^{-2}h^{-1}$ of H_2 production in the presence of visible light (100 $mWcm^{-2}$) [83]. Crystalline Fe_2O_3/TiO_2 nanocomposites have been prepared using Fe-containing nanoscale MOFs (metal organic frameworks) coated with amorphous TiO_2 followed by calcination. This material has been used for H_2 production from water by using visible light through photocatalysis [84]. Nano-CdS with a diameter of 2–3 nm is deposited onto the inner wall of TiO_2 nanotubes (TNTs) and then used for efficient photocatalytic H_2 production under visible light illumination. H_2 evolution through photocatalysis by a TNT-confined CdS nanocomposite (photostable) under visible light illumination is imputed due to the quantum size effect exhibited by nano-CdS, as a result of the spatial confinement effect of TNTs [85].

The mixed-phase of P25 TiO_2 with 85% anatase and 15% rutile has been used for H_2 production in the presence of alcohol/water systems. It was found that excited electrons across the rutile bandgap move to anatase lattice traps through interfacial surface sites, leading to a decrease in electron–hole pair recombination as well as an increase in charge carrier availability for photoreactions [86]. Nanoparticles of TiO_2 have been doped with a variety of materials, leading to different morphological, textural, and bandgap properties, and studied for their efficiency in H_2 production under visible light. In the report, two titanium sources, including titanium (IV) isopropoxide and titanium (IV) butoxide, were used for doping with different amounts of Zn (0%, 2%, or 5%) in the presence of an acidic solution (ascorbic acid, nitric acid) for a hydrolysis reaction at calcination temperatures of 500 °C and 600 °C simultaneously. It was reported that doped TiO_2 nanoparticles were able to produce H_2 efficiently in addition to degradation of methylene blue (under UV-light illumination or visible LED light). It was found that synthesized doped TiO_2 nanoparticles are porous aggregates having excellent crystallinity, mainly composed of an anatase phase [87].

Graphene-$Zn_xCd_{1-x}S$ nanocomposites have been prepared by a hydrothermal method using thiourea as an organic S source. Further, thiourea helps in situ growth of $Zn_xCd_{1-x}S$ nanoparticles on graphene nanosheets. It leads to intimate interfacial contact between graphene and $Zn_xCd_{1-x}S$ nanoparticles, which helps in effective transfer of photogenerated charge carriers, thus enhancing H_2 production through photocatalysis. The highest H_2 production rate has been found to be 1.06 mmoL $h^{-1}g^{-1}$, when graphene content is 0.5 wt% in graphene-$Zn_{0.5}Cd_{0.5}S$. The apparent quantum efficiency is 1.98% at 420 nm. Here, organic sources of S have been found to be much more effective than inorganic sources of S, like Na_2S, due to incomplete nucleation of graphene-$Zn_xCd_{1-x}S$ nanocomposite because of weak van der Waals forces [88]. ZnS nanoparticles based on the photosensitization of graphene, having a wide bandgap, are used for photocatalytic H_2 production from water splitting. Here, composites of ZnS nanoparticles and graphene (GR) sheets have been prepared using a two-step hydrothermal method based on $ZnCl_2$, Na_2S, and GO (graphite oxide) as starting materials. The prepared ZnS-GR can produce H_2 under visible light illumination.

ZnS nanoparticles have good interfacial contact with the 2D GR sheet, which helps in effective photocatalysis. Under optimized conditions, the content of GR in the catalyst is 0.1%, leading to maximum H_2 production of 7.42 μmoL $h^{-1} g^{-1}$ (eight times more than the pure ZnS sample) [89]. The nanoarchitectures of ZnS were fabricated with a homogenous dispersion of Au nanoparticles (5 nm) onto their surface. The addition of Au nanoparticles helped in narrowing the bandgap of ZnS. Optimal photocatalytic H_2 production of 3306 μmoL $h^{-1} g^{-1}$ takes place in the presence of sacrificial reagents like Na_2S and Na_2SO_3 under a 350 W xenon arc lamp, when Au content is 4%. Despite the presence of a strong surface plasmon resonance (SPR) absorption by Au nanoparticles on the surface of ZnS, no H_2 production has been observed in the visible light region with $\lambda > 420$ nm [90]. AgInZnS-reduced graphene (AIZS-rGO) nanocomposites with tunable bandgap absorption have been synthesized, leading to photocatalytic H_2 production under visible light irradiation. The AIZS-rGO nanocomposites with 0.02 wt% of graphene have a maximal production rate of H_2 of 1.871 mmoL $h^{-1} g^{-1}$ (nearly two times higher than pure AIZS nanoparticles). In the AIZS-rGO nanocomposites, graphene serves as a supporting layer and recombination center for conduction band electrons and valence band holes [91].

Nanoparticle alloy cocatalysts of platinum-palladium (Pt–Pd) loaded with CdS have been used for H_2 evolution under visible light illumination. It has been found that the composition and shape of nanoparticles are crucial in determining the efficiency of H_2 evolution. In the present report, two shapes of Pt–Pd nanoparticles are focused on: crystal planes {100} and nano-octahedra {111}. Pt–Pd nanocubes/CdS photocatalyst are 3.4 times more effective than Pt–Pd nano-octahedra/CdS with a Pt:Pd atomic ratio ranging from 1:0–2:1 leading to H_2 production ranging from 900 to 1837 μmoLh^{-1}. Most interestingly, the requirement of larger amounts of Pd (less costly than Pt) makes the process more cost effective [92]. Nanoferrite has been used as a catalyst for H_2 production through splitting of water in the presence of Pt as cocatalyst. The rate of H_2 evolution has been found to be 8275 μmoL $h^{-1} g^{-1}$ under illumination of visible light (much higher than that produced by commercial iron oxide, i.e., 0.0046 μmoL h^{-1}). Factors like configuration of the photoreactor, dosage of the photocatalyst, illumination intensity, irradiation time, sacrificial donor, and presence of a cocatalyst are crucial in determining the rate of H_2 evolution. For example, a slight alteration in the reactor configuration has increased H_2 evolution by sevenfold. Most importantly, temperature plays a significant role in determining the rate of H_2 evolution. Nanoferrites are extremely stable in producing H_2 up to 30 h with a cumulative H_2 evolution rate of 98.79 μmoL h^{-1} [93]. A novel 3D flower-like structure of nano-CdS has been prepared through a hydrothermal process by using $Cd(NO_3)_2 \bullet 4H_2O$ in the presence of thiourea and L-histidine. The photoelectrochemical performances of flower-like nano-CdS and pure CdS nanocrystals were compared with respect to H_2 production. A large difference was observed, with H_2 production rate being increased by 13-fold in the case of flower-like nano-CdS under visible light irradiation [94].

A very low-cost nanocatalyst for H_2 production based on molybdenum carbide (Mo_2C) nanoparticles supported on carbon nanotubes (CNTs) has been prepared and labeled as Mo_2C/CNT. The production of H_2 was reported in the temperature range 25–85 °C in the presence of 8 M KOH. Mo_2C/CNT has excellent stability with specific activity of 40 Ag^{-1} at −0.40 V [95]. MoO_{3-x} nanoparticles have been synthesized at room temperature using $(NH_4)_6 Mo_7O_{24}{\cdot}4H_2O$ and $MoCl_5$ as precursors. The size of the prepared MoO_{3-x} nanoparticles was found to be in the range of 90–180 nm (as reported by SEM and TEM images) with a molar ratio of Mo (VI)/Mo (V) of 1:1. These nanoparticles were found to show localized plasmon resonance (LSPR) properties as generated due to the presence of oxygen vacancies. With the presence of strong plasmonic absorption in the visible and near-infrared region, these nanostructures have excellent activity with respect to H_2 production towards visible light [96]. One interesting report discusses production of H_2 through splitting of sea water using nitrogen-doped, cobalt-encased CNTs in a cost-effective manner. It was reported that urea-derived CNTs at 900 °C have excellent catalytic properties with respect to H_2 production with operation stability from pH 1–14. These nanostructures, with high natural abundance, ease of synthesis, and high catalytic activity in addition to excellent durability in seawater, are potential candidates for industrial applications [97]. The most frequent problem seen with using nanoparticles for H_2 production is their high rate of aggregation during catalysis (water splitting), particularly with late transition metal nanoparticles. In one report, monodisperse metastable ruthenium nanoparticles (RuNPs) with a size of 5 nm were found to have minimal aggregating properties. It was found that these nanoparticles were stable without undergoing any aggregation even after 8 months at room temperature in aqueous media. The RuNPs have E_a of 27.5 kJ moL^{-1} with H_2 yield of 21.8 turnovers per min at 25 °C [98]. Quaternary $CuGa_2In_3S_8$ photocatalysts (particle size ≈4 nm) have been synthesized (one of the most tedious tasks) exhibiting n-type semiconductor characteristics with a transition bandgap of ≈1.8 eV. Under optimized conditions, these quaternary photocatalysts with 1.0 wt% Ru produced a maximum yield of H_2 with an apparent quantum efficiency of 6.9% ±0.5 at 560 nm [99]. H_2 produced using nanosilicon via water splitting in the absence of light, heat, or electricity has been reported for the first time. Here, the size of the nanosilicon is ~10 nm diameter; it reacts with water and generates 1000 times faster production of H_2 than ever reported by nanoparticles of Si with a size of 10 nm. This has been attributed to the drastic change in etching dynamics at nanoscale from anisotropic etching of larger silicon to effectively isotropic etching of 10 nm silicon [100].

2.5 Combo-technology (nanotechnology and enzyme technology) in production of clean energy

Nanoparticles have been very useful in utilization of renewable sources for various applications. It has been found that complementing nanotechnology with enzyme technology (which we here call *combo-technology*) can lead to powerful results in various respects, with easier commercialization and easier adaptability for common uses in daily life, thus leading more people to be involved in the utilization of renewable biomass resources and assisting in the ultimate complete replacement of fossil fuels in coming years.

Enzyme technology has been very helpful in removal of various toxic chemicals used in different processes related to renewable energy sources. Biofuels are predicted to be the fastest advancing research area to provide alternative energy sources that are both clean and sustainable. The following section provides a detailed report on the production of biofuels using combo-technology (Table 2.3).

2.5.1 Hydrogen production by combo-technology

The enzyme hydrogenase (H_2ase) with a [NiFeSe] cluster from *Desulfomicrobium baculatum* has been immobilized onto TiO_2 nanoparticles. The enzyme produces H_2 through a visible light-driven process with the help of dye-sensitized TiO_2 nanoparticles for effective light harvesting under ambient conditions. Here, the nanoparticles are Ru dye sensitized with triethanolamine as a sacrificial electron donor. Immobilized H_2ase can produce H_2 at a turnover frequency of ~50 (mol H_2 s^{-1}) (mol total hydrogenase) $^{-1}$ at pH7 and 25°C, under solar irradiation. This system has excellent electrocatalytic stability with the ability to work under anaerobic conditions as well as in conditions with prolonged exposure to air. This system has versatile applications in several different sectors due to its extensive robustness [101]. [NiFeSe]-based H_2ase has been immobilized onto TiO_2 nanoparticles modified with polyheptazine carbon nitride polymer, melon (CN_x). H_2 is produced through a light-driven process by solar AM 1.5G irradiation of immobilized H_2ase (having a turnover number of $>5.8\times10^5$ mol H_2 after 72h in a sacrificial electron donor solution). A quantum efficiency of 4.8% (photon-to-hydrogen conversion) has been achieved under irradiation with monochromatic light (UV-free solar light irradiation at $\lambda>420$nm). The CN_x-TiO_2-H_2ase immobilization system has set a novel benchmark for photocatalytic H_2 production using no toxic metal in the presence of visible light [102]. The evolution of H_2 by H_2ase from three different sources (*Clostridium pasteurianum, Desulfovibrio desulfuricans* strain Norway 4, and *D. baculatus* 9974) has been immobilized onto TiO_2 nanocrystals and compared. The evolution by immobilized H_2ase takes place through direct electron transfer in the presence of methanol/EDTA with efficiency of the process being enhanced by rhodium complexes under bandgap illumination at pH > 7.0. It was found that immobilized H_2ase from *C. pasteurianum* and *D. baculatus* 9974 are much more efficient in producing H_2 in the presence of methanol. Further, rhodium tris- and bis-bipyridyl complexes can act as efficient electron carriers when there is inefficient direct transfer [103].

Hybrid complexes of CdTe nanocrystals (nc-CdTe) have been used as an immobilizing matrix for H_2ase from *Clostridium acetobutylicum* having an [Fe-Fe] cluster. The enzyme has been immobilized through electrostatic interactions to form an immobilized enzyme having exceptionally good stability. Photoproduction of H_2 is the function amount of H_2ase and nc-CdTe taken for immobilization. It was reported that when H_2ase concentration was lower than the nc-CdTe concentration during immobilization, it resulted in a maximal rate of H_2 photo-production due to enhanced

Table 2.3 Enzymes immobilized onto various nanomaterials for efficient energy production from renewable sources.

Hydrogenase	Lipase	Cellulase	Invertase	β-glucosidase	Xylanase
TiO_2 NPs	$BaFe_{12}O_{19}$ NPs	$Fe_3O_4.SiO_2$ NPs	Fe_3O_4/SiO_2 NPs	SiO_2 NPs	Silicon oxide NPs
TiO_2/polyheptazine carbon nitride polymer NPs	Fe_2O_3 NPs	Chitosan-coated magnetic NPs	Magnetic diatomaceous earth NPs	Fe_3O_4 NPs coupled with agarose chelated with Cu^{2+}, Zn^{2+}, Cr^{2+}, Ni^2, Co^{2+}	Fe_2O_3/ hyperbranched polyglycerol NPs
TiO_2 NCs	Mesoporous SiO_2 NPs	Graphene oxide nanosheets	Chitosan/γ-Fe_2O_3 NPs	Fe_3O_4/PMG/IDA-Ni^{2+} NPs	Silicon oxide NPs
CdTe NCs	Wrinkled SiO_2 NPs	FeO NPs	Cu^{2+}, Zn^{2+}, Cr^{2+}, Ni^{2+} are chelated onto polyvinylimidazole-grafted Fe_2O_3 NPs	Fe_2O_3 NPs	
CdS NRs	SWCNTs	MgO NPs	Chitosan NPs	Silicon oxide NPs	
Ni/heptazine carbon nitride polymer NPs	MWCNTs	TiO_2 NPs		Fe_2O_3/SWCNTs	
Nanolipoprotein	Mesoporous SiO_2/ FeO NPs				
	Fe_2O_3 NPs				
	Fe_3O_4-SiO_2 NPs				
	Ag NPs				

electron transfer. The extent to which the intramolecular electron transfer (ET) is dependent on the ratio of intrinsic radiative and nonradiative (heat dissipation and surface trapping) recombination pathways as followed by photoexcited nc-CdTe. Further, the duration of H_2 photo-production depends on the stability of nc-CdTe under the reaction conditions. Under optimized conditions, photon-to-H_2 efficiency has been found to be 9% under monochromatic light and 1.8% under AM 1.5 white light, using ascorbic acid as a sacrificial electron donor [104]. CdS nanorods have been prepared and capped with 3-mercaptopropionic acid (MPA) and used as an immobilizing matrix for H_2ase with [Fe-Fe] cluster from *Clostridium acetobutylicum.* H_2 has been produced by immobilized enzyme by the photochemical process with reduction of H^+ to H_2 at a turnover frequency of 380–900 s^{-1} with a photoconversion efficiency of 20% by illumination at 405 nm in the presence of a sacrificial donor. Factors like the amount of enzyme to nanoparticles, sacrificial donor concentration, and light intensity on photocatalytic H_2 production are the determining factors for electron transfer, hole transfer, or rate of photon absorption, respectively. Further, photoproduction of H_2 was found to be similar by using monochromatic light at 405 nm or by AM 1.5 solar fluxes. Immobilized enzyme can produce H_2 for up to 4 h with a total turnover number of 10^6 before photocatalytic activity is completely lost due to inactivation caused due to photo-oxidation of the immobilizing matrix CdS capped with MPA [105].

H_2 has been produced by photolysis of water in the presence of solar light in water by using immobilized H_2ase with a [NiFeSe] cluster onto bioinspired synthetic nickel nanocatalyst (NiP) containing heptazine carbon nitride polymer, and melon (CN_x) on its surface. It was found that immobilized H_2ase has shown turnover numbers of >50,000 mol H_2 and ~155 mol H_2 in redox-mediator-free aqueous solution at pH 6.0 and 4.5, respectively, under UV-free solar light irradiation ($\lambda > 420$ nm) [106]. A novel immobilized H_2ase has been prepared using nanolipoprotein particles (NLPs). NLPs, formed from apolipoproteins and phospholipids, provide a well-defined water-soluble environment to immobilized H_2ase and maintain its enzymatic activity. H_2 production can take place in an oxygen-restricted environment with excellent turnover numbers [107].

2.5.2 Biodiesel production by combo-technology

Lipases are excellent biocatalysts in the production of biodiesel due to their remarkable biochemical and physiological properties. They are also used in various other industrial applications for alcoholysis, acidolysis, aminolysis, and hydrolysis reactions. The production of biodiesel by lipase was first reported by Mittelbach [108]. Lipase follows the ping-ping bi-bi mechanism in undergoing transesterification, which involves two steps: hydrolysis of the ester bond followed by esterification with the second substrate. Lipases are available from various sources ranging from plants (papaya latex, oat seed lipase, and castor seed), animals (pig and human pancrease), bacteria (*Chromobacterium viscosum, Pseudomonas cepacia, Pseudomonas fluorescens, Photobacterium lipolyticum Streptomyces sp.,* and *Thermomyces lanuginose*), filamentous fungi (*Aspergillus niger, Mucor miehei, Rhizopus oryzae*) and yeast (*Candida antarctica, Candida rugosa*). However, yeast is most commonly used for biodiesel production. Lipases from microorganisms are much preferred at industrial scale for the production of biodiesel due to their short generation time, in addition to high yield of product (biodiesel), great versatility when faced with extreme environmental conditions, and simplicity in genetic manipulation and cultivation conditions. The enzyme lipase has been categorized into three groups based on its specificity: 1, 3-specific lipases, fatty acid–specific lipases, and nonspecific lipases. Mostly 1, 3-specific lipases are used for biodiesel production, involving release of fatty acids from positions 1 and 3 of a glyceride followed by hydrolysis of ester bonds in those positions. Immobilized lipase (attaching enzyme onto insoluble solid support) is more preferable for industrial processes, as it is economical, selective, cost effective (due to reusability), and stable. Nanoparticles are discussed in the following paragraphs for their utility as an immobilizing matrix for lipase immobilization.

Lipase from *Rhizomucor miehei* (RML) has been immobilized through formation of hybrid magnetic cross-linked lipase aggregates. These surfactant-activated magnetic RML cross-linked enzyme aggregates (CLEAs) have excellent recovery of 2058% with increment of kinetic parameters by 20-fold with respect to soluble enzymes. The immobilized RML in the form of CLEAs has produced biodiesel with a yield of 93%. Immobilized enzyme can be easily separated from the reaction mixture through simple magnetic decantation. Immobilized enzyme can be reused five times with retention of 84% initial activity [109]. Lipase has been produced by gamma-irradiated *Aspergillus niger* ADM110 fungi and immobilized onto magnetic barium ferrite nanoparticles (BFNs) and used for the production of biodiesel. Biodiesel has been produced from waste cooking oil (WCO) at 45 °C, with incubation time of 4 h and rotation speed of 400 rpm. Parameters like flashpoint, calorific value, and cetane number are found to be 188 °C, 43.1 MJ kg^{-1} and 59.5, respectively. The acid values of WCO and FAMEs are reported as 1.90 and 0.182 (mg KOH g^{-1} oil), respectively. Immobilized lipase can be reused for five cycles with retention of 87% of initial activity [110]. Most interestingly, lipase from *Thermomyces lanuginosus* has been coated with Fe_2O_3 nanoparticles through precipitation. Here biodiesel has been synthesized by the immobilized lipase in a solvent-free medium by the alcoholysis of oils/fats like soybean oil. Optimized immobilized enzyme can produce biodiesel in 3 h with oil:ethanol (*w*/w) of 1:4 at 40 °C in the presence of 20% (w/w) silica (facilitate acyl migration), resulting in the conversion of oil to biodiesel by 96% [111].

Mesoporous silica nanoparticles that have been prepared using tannic acid (TA-MSNs) are monodisperse spherical particles with an average diameter of 195 ± 16 nm. They have a surface area of 447 $m^2 g^{-1}$, a large pore volume of 0.91 $cm^3 g^{-1}$, and mean pore size of 10.1 nm as found from BET (Brunauer–Emmett–Teller). These nanoparticles have been used for immobilization of lipase from *Burkholderia cepacia* through physical adsorption. Immobilized lipase has good thermostability with tolerance towards a wide range of organic solvents like methanol, ethanol, isooctane, n-hexane, and tetrahydrofuran. Immobilized enzyme has excellent operational reusability in esterification and transesterification reactions, with recycling of 15 times.

Total biodiesel yield by optimized immobilized enzyme has been reported to be 85% [112]. Wrinkled silica nanoparticles (WSNs) are prepared with highly ordered structure and radially oriented mesochannels by the solvothermal method. These nanoparticles have particle size ranging from 240 to 540 nm, specific surface area from 490 to 634 $m^2 g^{-1}$ with a unique surface morphology characterized as radial wrinkled structures. These nanoparticles have been used as an immobilizing matrix for lipase from *Candida rugosa* (CRL) and used for esterification of oleic acid with methanol to produce biodiesel. Under optimized conditions, the highest oleic acid conversion rate achieved was about 86.4% [113].

Candida antarctica lipase B (CaLB) has been immobilized onto single-walled carbon nanotubes (SWCNTs) via the covalent method and used in biodiesel production with sunflower oil. The obtained immobilized CaLB had minimal diffusional limitations as observed in batch mode and was found to be highly efficient as well as stable in acetonitrile. The immobilized enzyme has produced biodiesel with a yield of 83.4% after an incubation time of 4 h at 35 °C. Immobilized CaLB has a reusability of 10 cycles with a retention of >90% of original activity [114]. MWCNTs have been surface modified, particularly with carboxyl groups, and called MWCNT-COOH. These modified MWCNTs have been used for immobilizing CALB. Functionalized MWCNT-COOH has been further modified by amidation with either butylamine (MWCNT-BA) or octadecylamine (MWCNT-OA) for effective immobilization of CALB. Immobilized CALB has been used for catalyzing methanolysis of rapeseed oil to produce biodiesel. It has been found that higher enzyme loading is required when MWCNT-BA has been used for immobilization, i.e., 20 mg protein g^{-1} in the case of MWCNT-BA and 11 mg protein g^{-1} in the case of MWCNT-OA. Comparative analyses on the yield of biodiesel by CALB immobilized onto MWCNT-BA and MWCNT-OA have found that the former has a higher yield of 92% while the latter has a yield of 86%. Immobilized lipase is thermostable with good catalytic capability in repeated batch experiments [115]. CLEAs of CALB (*Candida antarctica* lipase B) have been prepared by cross-linking enzyme covalently with magnetic nanoparticles that have their surface functionalized with $-NH_2$ groups. CLEAs of CALB have been used for the synthesis of biodiesel from inedible vegetables (unrefined soybean, jatropha, cameline) and waste frying oils. It was reported that 1% mCLEAs (w/w of oil) can produce biodiesel with a yield of 80% at 30 °C after 24 h of reaction, and 92% at 30 °C after 72 h. CLEAs of CALB have excellent stability as well as feasibility of re-utilization after 10 cycles without any apparent loss in activity. CLEAs of CALB can be easily removed from a reaction mixture after their catalysis due to their magnetic character. They are strongly recommended for industrial applications [116].

Lipase has been immobilized onto mesoporous silica/iron oxide magnetic core shell nanoparticles by using optimization through response surface methodology (RSM). Immobilized lipase has been used in the production of biodiesel by transesterification of waste cooking oil (WCO) to partially substitute for petroleum diesel. RSM has also been used in optimizing various parameters involved in the synthesis of petroleum diesel. A quadratic response surface equation fits well for calculating FAME (fatty acid methyl ester, i.e., diesel) content based on experimental data in accordance with central composite design. As per the RSM-based model, predicted maximum FAME content has been found to be 91% at optimum level of variables including: methanol ratio to WCO (4.34), lipase content (43.6%), water content (10.22%), and reaction time (6h). The immobilized lipase can be used four times without any apparent loss in activity [117]. The lipase from *Pseudomonas cepacia* has been immobilized onto MNPs and used for biodiesel production from WCO. The maximum production of biodiesel can be obtained through optimization as done by RSM. The optimal parameters as obtained are: dosage of immobilized lipase is 40% (w/w of oil), reaction temperature is 44.2 °C, substrate molar ratio is 5.2, and water content is 12.5%. Methanol has been added stepwise at an interval of 12h. The biodiesel yields as predicted from RSM and experimental data are found to be 80% and 79%, respectively [118]. Two lipases from different sources, *Thermomyces lanuginosus* (TLL) and *Candida antarctica* (CALB), are taken and immobilized covalently onto core-shell structured Fe_2O_3 magnetic nanoparticles with size of 80nm, followed by freeze drying to obtain magnetic nanobiocatalyst aggregates (MNAs). These synthesized MNAs have sizes ranging from 13 to 17 μm with enzyme loading of 61 mg TLL or 22 mg CALB per gram MNA. These MNAs are used for producing biodiesel from grease with a yield range of 95%–99%. MNA TLL has been found to have better performance than MNA CALB, which has a yield of 99% in 12h using 3.3wt% catalyst, while the latter has a yield of 97% in 12h using 0.45% wt of catalyst from grease (17% FFA) in the presence of methanol [119].

Lipase has been produced from an isolated strain *Burkholderia* sp. C20 and immobilized onto core shell nanoparticles prepared by coating Fe_3O_4 core with silica shell followed by treatment with dimethyl octadecyl [3-(trimethoxysilyl) propyl] ammonium chloride. The immobilized lipase has an efficiency of 97% in the case of alkyl-functionalized Fe_3O_4-SiO_2, while it is 76% in the case of nonmodified Fe_3O_4-SiO_2. The kinetic studies on immobilized lipase revealed that the Michaelis constant (K_m) and maximum reaction rate velocity (V_{max}) were found to be 6251 U g^{-1} and 3.65 mM, respectively. The immobilized lipase can catalyze the transesterification of olive oil with methanol to produce biodiesel with a conversion rate of over 90% within 30h in a batch operation by using 11 wt% immobilized lipase. The immobilized lipase has a reusability of 10cycles without any catalytic loss during transesterification [120]. Commercially available lipase has been immobilized onto silver nanoparticles with a size range of 10–20nm coated with polydopamine (immobilized lipase/PD/AgNPs complex labeled as LPA). LPA has been used for the production of biodiesel from soybean oil with a yield of 95% at 40 °C for 6h of reaction time, while the yield reduced drastically to 86% when lyophilized lipase was used. The LPA has reusability of 7cycles and with the latter cycles, biodiesel yield was decreased to 27% [121].

2.5.3 Bioethanol production by combo-technology

The production process for bioethanol using lignocellulosic materials makes use of enzymes like cellulase, invertase, β-glucosidase, and xylanase. Lignicellulosic biomass is obtained in abundance, because cellulose and the hemicelluloses present in it are the basic cross-linked structures found in the plant cell wall. Previously, starch grains (main energy source in livestock feedstock) were used, but this has led to a significant increase in their cost, making it unfit to use for biofuel production. The bioethanol production from cellulosic biomass involves biomass pretreatment, hydrolyzation of biomass into fermentable sugar, followed by ethanol production from reducing sugar. The enzymes such as cellulase, invertase, β-glucosidase, and xylanase are used in the first step, leading to complete replacement of the acid used previously, which was toxic, time consuming, and nonspecific. Enzyme-based hydrolysis can be carried out in mild conditions, leading to high production of pure sugar without generating any toxic wastes. Further, the process is cost effective and not laborious, making it fit for industrial applications. The following paragraphs discuss utilization of nanoparticles for immobilization of enzymes involved in hydrolysis of lignocellulose and other agricultural residues, imparting better stability, reusability, resistance towards extreme physicochemical conditions, as well as better catalysis.

Cellulases from *Trichoderma reesei* have been immobilized onto chitosan-coated magnetic nanoparticles, using glutaraldehyde as coupling agent, and used for lignocellulosic material saccharification for bioethanol production. The average diameters of synthesized chitosan-coated magnetic nanoparticles before and after enzyme immobilization have been found to be 8 and 10 nm, respectively. The immobilized enzyme has a K_m value increased by 8 times with retention of 37% of initial soluble enzyme activity after immobilization. Immobilized cellulase can be used for about 15 cycles with retention of 80% of activity with respect to the first cycle and it bears better thermal and storage stability with respect to soluble enzymes [122]. Supermagnetic $Fe_3O_4.SiO_2$ nanoparticles are prepared having a core-shell structure with a diameter of 10 nm. These nanoparticles are molecular-imprinted (creating template shaped cavities) for specific cellulase adsorption. An excellent immobilization efficiency of 95% has been achieved after optimization. Immobilized cellulase has an increased shelf half-life by 3.3-fold with respect to soluble enzyme at 70 °C. Further, immobilized cellulase has the same optimal pH with higher optimal temperature, better thermal stability, and higher catalytic efficiency with respect to soluble enzyme.

Immobilized cellulase has been used not only for bioethanol production but also finds applications in the paper and pulp industry, as well as in the pharmaceutical industry [123]. Cellulase has been immobilized onto PEGylated graphene oxide (GO) nanosheets for saccharification of lignocelluloses with retention of 61% of the initial activity with respect to soluble enzyme. The stability of immobilized cellulase has been enhanced by 30 times with a K_m of 3.2 mg mL^{-1} (K_m of soluble cellulase is 2.7 mg mL^{-1}) [124]. A novel magnetic cross-linked aggregate (CLEAs) of cellulase has been developed for biomass bioconversion into biofuel. These CLEAs of cellulase have wider pH and temperature with respect to soluble enzyme, with a recyclability of 6 in addition to a retention of 74% after immobilization. They can be easily separated from the reaction mixture by applying a magnetic field. They can hydrolyze bamboo biomass with a yield of 21% and are shown to have good potential for biomass applications [125].

Superparamagnetic FeO nanoparticles have been used as an immobilizing matrix for cellulase and were found to provide stable enzyme bonding and excellent colloidal stability. Here cellulase was loaded, with 0.43 g per g of matrix, leading to a recovery of 75% with recyclability of 10 times. Further, it was reported that enzymatic activity is well preserved with outstanding lifecycle stability [126]. The most common problems with using magnetic nanoparticles for enzyme immobilization are the tendency of aggregation, with higher reactivity, thus higher susceptibility towards oxidation by air. This can be solved by encapsulating magnetic nanoparticles inside polymeric nanospheres and then using them for enzyme immobilization. Cellulase has been immobilized onto encapsulated magnetic nanoparticles with an average diameter of 150 nm. Immobilized cellulase has an optimum temperature in the broader range with exceptionally good reusability capacity, with retention of 69% of the initial enzyme activity, even after eight cycles of usage. Immobilized enzyme can be easily separated from the reaction mixture by applying a single magnet [127]. Cellulase from a psychrophilic strain of *Bacillus subtilis* has been immobilized onto MgO nanoparticles, followed by linking to graphene oxide nanosupport via glutaraldehyde. It was reported that immobilized cellulase has increased enzymatic activity by 2.98-fold with respect to soluble cellulase at 8 °C. Further, immobilized cellulase has reduced K_m by 6.7-fold while V_{max} increased by 5-fold at 8 °C. Immobilized cellulase can work from 5 °C to 90 °C with little effect on enzymatic activity. Immobilized cellulase has retained enzymatic activity even after 12 repeated uses with storage stability of 120 days at 4 °C. Comparative studies on soluble and immobilized cellulase have revealed that immobilized enzyme has $t_{1/2}$ (half life) and E_d increased by 72.5 times and 2.48 times respectively at 90 °C, while the numbers are 41.6 times and 2.19 times respectively at 8 °C [128].

Cellulase from *Aspergillus niger* has been immobilized onto TiO_2 nanoparticles by two methods: physical adsorption and covalent coupling. In the case of the covalent method, the enzyme was bonded after modification of TiO_2 nanoparticles by aminopropyltriethoxysilane (APTS). Comparative analyses of the two approaches have found that the adsorbed and covalently immobilized enzymes have 76% and 93% of enzymatic activity, respectively, with respect to a soluble enzyme. The catalytic efficiency (V_{max}/K_m) was increased from 0.4 to 4.0 in the case of covalent attachment, while there was a very slight increase from 0.4 to 1.2 in the case of the adsorbed enzyme. The covalently attached enzyme had better thermal and operational stability as well as reusability with respect to the adsorbed enzyme. Studies by HR-TEM (high-resolution transmission electron microscopy) and AFM (atomic force microscopy) found that the enzyme was severely aggregated in the case of immobilization by adsorption, while there was an even monolayer of enzyme on the matrix in the case of the covalent mode of attachment. The lower amount of enzyme activity and thermal stability has been attributed to severe enzyme aggregation during enzyme immobilization by adsorption [129].

The enzyme invertase has been immobilized onto magnetic diatomaceous earth nanoparticles (mDE-APTES-invertase). Immobilized enzyme has excellent sucrolytic activity obtainable through an easier and lower-cost method. Characterization of immobilized invertase has revealed that residual specific activity of immobilized enzyme is 92.5%, with specific activity increased by 2.42 times with respect to the soluble enzyme. Thermal and storage stability for immobilized invertase are found to have a retention of 85% activity after 60 min at 35 °C with a storage period of 120 days (retention of 80% of enzymatic activity), respectively [130]. Magnetic nanoparticles (Fe_3O_4) have been modified with silica ($Fe_3O_4.SiO_2$) followed by adding glycidylmethacrylate (GMA) [surface initiation] and then attaching a spacer arm (hexamethylene diamine) and labeling the result as $Fe_3O_4.SiO_2$.pGMA-SA-3. Enzyme invertase has been immobilized onto a fabricated nanoparticle matrix with the amount as $33.4 \pm 1.3\,mg\,g^{-1}$. Kinetic values (K_m and V_{max}) of immobilized invertase are: $39.4\,mmol\,L^{-1}$ and 349.5 $mmoL\,L^{-1}\,min^{-1}$ (slight change with respect to soluble enzyme having K_m and V_{max} as $34.3\,mmol\,L^{-1}$ and $387.2\,mmol\,L^{-1}\,min^{-1}$, respectively) [131]. Chitosan-coated sol–gel derived γ-Fe_2O_3 magnetic nanoparticles (MNPs) are used as an immobilizing matrix for invertase. Immobilized invertases onto MNPs (IIMNPs) are more stable at varying pH and temperature conditions, with reusability of 20 times without significant loss in enzymatic activity [132]. Various metal ions (Cu^{2+}, Zn^{2+}, Cr^{2+}, Ni^{2+}) are chelated onto polyvinylimidazole-grafted Fe_2O_3 magnetic nanoparticles (PVIgMNPs) and then used for immobilization of invertase via adsorption. The maximum invertase immobilization capacity has been found in the case of Cu^{2+} chelation, with the value corresponding to $142.856\,mg\,g^{-1}$ at pH 5.0 [133]. Invertase from *Saccharomyces cerevisiae* has been covalently immobilized onto chitosan nanoparticles (activated with glutaraldehyde). Various parameters like thermal and storage stabilities, optimal pH, and temperature of immobilized enzyme are found to be unaltered with respect to the soluble enzyme. Further, the immobilized enzyme can be reused for 59 batches without any loss in enzymatic activity and with excellent operational stability, with respect to various reports published so far on immobilized invertase [134].

Enzyme β-glucosidase (BGL) is immobilized onto SiO_2 nanoparticles with immobilization efficiency of 52% with a yield of 14.1%. Optimal temperature and pH of immobilized enzyme are: 60 °C, pH 5.0 with stability in temperature range from 60 to 70 °C. Kinetic values (K_m and V_{max}) of immobilized enzyme are found to be: 1.074 mM (K_m of soluble enzyme is 0.9 mM), $1.513\,U\,mg^{-1}$ (V_{max} of soluble enzyme is $3.5\,U\,mg^{-1}$). The immobilized enzyme has showed improved storage stability at temperatures 4 and 25 °C and with reusability of up to 10 cycles with 70% residual activity. Immobilized BGL can easily act on a variety of sources like sugarcane juice, caffeic acid, etc. [135]. Magnetic Fe_3O_4 nanoparticles coupled with agarose (AMNPs) have been chelated with different metal ions (Cu^{2+}, Zn^{2+}, Cr^{2+}, Ni^{2}, Co^{2+}) and used for BGL immobilization. It was observed that maximum immobilization efficiency was found corresponding to Co^{2+}-chelated AMNPs with enzyme adsorption capacity of $1.81\,mg\,g^{-1}$ particles. The Michaelis constant (K_m) and V_{max} of the immobilized BGL were 0.904 mM and $0.057\,\mu mol\,min^{-1}$, respectively, with activation energy being much lower with respect to soluble enzyme. The immobilized BGL has exhibited improved thermostability and operational stability with retention of >90% of its initial enzyme activity after 15 successive batches. The results so obtained confirmed usage of immobilized BGL for various industrial applications [136].

The BGL gene from *Coptotermes formosanus* has been expressed in *E. coli* BL21. The enzyme produced is purified and immobilized onto Fe_3O_4/PMG/IDA-Ni^{2+} nanoparticles [Fe_3O_4/PMG (poly (*N, N′*-methylenebisacrylamide-co-glycidyl methacrylate) core/shell microspheres with iminodiacetic acid (IDA), which is used to open epoxy rings present on the shell of microspheres to the combination of Ni^{2+}]. BGL has been attached onto the surface of Fe_3O_4/PMG/IDA-Ni^{2+} to form Fe_3O_4/PMG/IDA-BGL via covalent binding through an imidazolyl link with Ni^{2+}. The immobilized BGL has excellent catalytic activity and stability with respect to soluble enzyme with reusability of 30 cycles with retention of 65% of the original activity [137]. The thermostable BGL from *Aspergillus niger* has been immobilized onto functionalized magnetic nanoparticles using covalent bonding. Immobilized BGL has 93% of immobilization efficiency with pH optima at 6.0 (soluble enzyme pH optima at 4.0) and temperature optima at 60 °C (similar to soluble enzyme). The Michaelis constants (K_m) of immobilized and soluble enzymes are 3.5 and 4.3 mM, respectively. Immobilized BGL has enhanced thermal stability at 70 °C with reusability of 16 cycles with retention of 50% of activity [138]. Crude enzyme preparation of BGL from *Agaricus arvensis* has been covalently immobilized onto functionalized silicon oxide nanoparticles with an immobilization efficiency of 158%. Thermostable immobilized BGL has a half-life increased by 288-fold at 65 °C with apparent V_{max} and k_{cat} as 3028 U mg protein^{-1} and 4945 s^{-1} (V_{max} and k_{cat} of soluble enzyme are: 3347 U mg protein^{-1}, 5466 s^{-1}, respectively). The immobilized enzyme retained 95% of the original activity after 25 cycles, and is thus an appropriate candidate for commercial applications [139]. Fe_2O_3 has been incorporated into single-walled carbon nanotubes to produce magnetic SWCNTs (mSWCNTs) and used as an immobilizing matrix for amyloglucosidase (AMG). AMG has been immobilized onto mSWCNTs via two approaches: physical adsorption and covalent immobilization, with the latter found to be more useful. Immobilized enzyme has a catalytic efficiency of 40% as found by using starch hydrolysis with reusability of 10 cycles. The factors like enzyme loading, activity, and structural changes after immobilization onto mSWCNTs have been thoroughly studied. Immobilized enzyme has retained complete activity when stored at 4 °C for at least 1 month. The properties of immobilized enzyme combined with the unique intrinsic properties of nanotubes have opened a path towards implementation in large-scale bioreactors [140].

Purified *Armillaria gemina* xylanase (AgXyl), having a catalytic efficiency $k_{cat}/K_m = 1440$ mg mL^{-1} s^{-1}, has been immobilized onto functionalized silicon oxide nanoparticles with an immobilization efficiency of 117%. AgXyl immobilization has led to a shift in the optimal pH and temperature with respect to soluble enzyme with a fourfold increase in half life. Immobilized AgXyl has a reusability of 17 cycles with a retention of 92% of the original enzyme activity [141]. A novel immobilizing matrix is based on magnetic nanoparticles supported by hyperbranched polyglycerol (MNP/HPG) and a derivative conjugated with citric acid and labeled as MNP/HPG-CA for immobilization of xylanase. Immobilized xylanase has exhibited substantial enhancement in reactivity, reusability, and stability. This novel immobilizing matrix has been recommended for various industrial enzymes due to its excellent properties [142]. Purified xylanase from *Pholiota adipose* has been reported to have the highest k_{cat} value ($4261\,s^{-1}$) reported so far. This enzyme has been immobilized onto nanoparticles of silicon oxide with immobilization efficiency of 66% with excellent hydrolytic activity towards xylooligosaccharides. The immobilized enzyme has a reusability of 17 cycles with retention of 97% of the original enzymatic activity and fits well for industrial applications in hydrolysis of xylooligosaccharides [143].

2.6 An outlook: a future roadmap to renewable and sustainable energy based on combining nanotechnology and enzyme technology

The energy crisis has been a serious issue around the globe due to limited sources of fossil fuels. Besides the energy crisis, serious concerns about rising air pollution and emissions of greenhouse gases have created an alarming situation for all countries regardless of their economic status. There is an urgent need to find sustainable energy sources, particularly renewable sources that are infinite in supply, in addition to being both clean and green in all respects. Several technologies have been developed to trap maximum energy from renewable sources and use it for various applications. Most importantly, sectors like residential, industrial, and transportation are the major energy-consuming arenas worldwide. Germany is the leading country in the world in this area, with the largest economy based on renewable sources of energy. It has been estimated that a 1% increase in renewable energy consumption has led to economic growth in Germany of 0.2194% [144]. The United States has been actively involved in promoting maximum usage of renewable sources in various applications [145]. Many developing countries like Thailand, South Africa, Egypt, Nigeria, Mali, and Ghana are actively involved in initiating energy saving, conservation, and management plans [146–148]. Further, several programs have been started worldwide for promoting active technology transfer for more usage of renewable sources [149].

Our most pressing need in the present scenario is to deploy various renewable energy sources in a green manner, thus meeting our energy needs as well as keeping our environment safe and healthy for future generations [150, 151]. Systematic research has been ongoing in various areas, with several promising claims. In the present context, technologies like nanotechnology and enzyme technology are helpful individually and they also complement each other in combination. They involve researchers from different fields, providing the additional advantage of having different points of view for implementation of renewable sources. A crucial requirement is that any technology adopted for use with renewable energy sources should be such that it can completely supersede nonrenewable energy sources like coal, oil, and natural gas.

Further, there should be various energy conservation awareness campaigns initiated at the government level as well as by local engineers and scientists, to make people aware of the significance of energy conservation and alternative energy technologies. There should be strengthening deterrents towards the use of fossil fuels, due to their evident harm to the environment and living beings. Governments should revise their energy policies to cope with the energy crisis by making full use of renewable energy sources. There should be no restraints in technology-exchange programs, so that developing and undeveloped countries can easily establish, build, and reinforce the renewable energy sector with the help of developed countries. More and more business people should invest in the renewable energy sector to generate more employment opportunities, besides helping to solve the energy crisis. Social media should be actively involved in educating people about energy sources and their utilization. Further, schools, colleges, and universities should have a compulsory course on energy conservation and utilization. Experts in various domains should be actively involved in seeking innovative solutions, which would be helpful in resolving the energy crisis more rapidly.

References

[1] Soytas U, Sari R. Energy Econ 2003;25:33–7.
[2] Laurmann JA. Clim Change 1985;7:261–5.
[3] Wigley TML, Jones PD, Kelly PM. Nature 1981;291:285.
[4] Kakinami Y, Kamogawa M, Liu JY, Watanabe S, Mogi T. Adv Space Res 2011;48:1613–6.
[5] Chino M, Nakayama H, Nagai H, Terada H, Katata G, Yamazawa H. J Nucl Sci Technol 2011;48:1129–34.
[6] Popp M. Science 1982;218:1280–5.
[7] He L, Ding Z, Yin F, Wu M. Springerplus 2016;5:1001.
[8] Moustafa K. Sci Total Environ 2017;598:639–46.
[9] Ramanathan V, Feng Y. Atmos Environ 2009;43:37–50.
[10] Hu Y, Cheng H. Nat Commun 2016;8:1–9.
[11] Jorgenson AK. Nat Clim Change 2012;2:398–9.
[12] Huang D, Zhou H, Lin L. Energy Procedia 2012;16:1874–85.
[13] Datta A, Mandal BK. Energy Technol Policy 2014;1:8–14.
[14] Moore RH, Thornhill KL, Weinzierl B, Sauer D, D'Ascoli E, Kim J, Lichtenstern M, Scheibe M, Beaton B, Beyersdorf AJ, Barrick J, Bulzan D, Corr CA, Crosbie E, Jurkat T, Martin R, Riddick D, Shook M, Slover G, Voigt C, White R, Winstead E, Yasky R, Ziemba LD, Brown A, Schlager H, Anderson BE. Nature 2017;543:411–5.
[15] Mushrush GW, Wynne JH, Willauer HD, Lloyd CL. J Environ Sci Health A Tox Hazard Subst Environ Eng 2006;41:2495–502.
[16] Wackett LP. J Microbial Biotechnol 2012;5:450–1.
[17] Tsai JH, Chen SJ, Huang KL, Lin WY, Lee WJ, Lin CC, Hsieh LT, Chiu JY, Kuo WC. Sci Total Environ 2014;466:195–202.
[18] Lopes ML, de Lima Paulillo SC, Godoy A, Cherubin RA, Lorenzi MS, Giometti FHC, Bernardino CD, de Amorim Neto HB, de Amorim HV. Braz J Microbiol 2016;47:64–76.
[19] Tenenbaum DJ. Environ Health Perspect 2008;116:254–7.
[20] MacLean HL, Lave LB, Lankey R, Joshi SJ. Air Waste Manag Assoc 2000;50:1769–79.
[21] Metje M, Frenzel P. Appl Environ Microbiol 2005;71:8191–200.
[22] Kim I, Han OH, Chae SA, Paik Y, Kwon SH, Lee KS, Sung YE, Kim H. Angew Chem Int Ed Engl 2011;50:2270–4.
[23] Rhodes CJ. Sci Prog 2010;93:37–112.
[24] Inganäs O, Sundström V. Ambio 2016;45:15–23.
[25] Zahedi A. Renew Energ 2006;31:711–8.
[26] Barron-Gafford GA, Minor RL, Allen NA, Cronin AD, Brooks AE, Pavao-Zuckerman MA. Sci Rep 2016;6:35070.
[27] Gulagi A, Choudhary P, Bogdanov D, Breyer C. PLoS One 2017;12:e0180611.

[28] Humada AM, Hojabri M, Sulaiman MHB, Hamada HM, Ahmed MN. PLoS One 2016;11:e0152766.
[29] Winsche WE, Hoffman KC, Salzano FJ. Science 1973;180:1325–32.
[30] Boz N, Degirmenbasi N, Kalyon DM. Appl Catal B 2009;89:590–6.
[31] Tang Y, Ren HM, Chang FQ, Gu XF, Zhang J. RSC Adv 2017;7:5694–700.
[32] Mahdavi M, Abedini E, Darabi AH. RSC Adv 2015;5:55027–32.
[33] Feyzi M, Shahbazi Z. J Taiwan *Inst Chem* Eng 2017;71:145–55.
[34] Amini G, Najafpour GD, Rabiee SM, Ghoreyshi AA. Dent Tech 2013;36:1708–12.
[35] Tahvildari K, Anaraki YN, Fazaeli R, Mirpanji S, Delrish E. J Environ Health Sci Eng 2015;13:1–9.
[36] Bala VSS, Kumar PS, Varathachary TK, Kirupha SD, Sivanesan S. Energy sources. Part *A,* recovery. Util Environ *Effects* 2016;38:2104–10.
[37] Pandit PR, Fulekar MH. J Environ Manage 2017;198:319–29.
[38] Teo SH, Islam A, Masoumi HRF, Taufiq-Yap YH, Janaun J, Chan ES, Khaleque MA. Renew Energ 2017;111:892–905.
[39] Kaur M, Ali A. Renew Energ 2011;36:2866–71.
[40] Kaur N, Ali A. Energy sources. Part *A,* recovery. Util Environ *Effects* 2013;35:184–92.
[41] Feyzi M, Norouzi L. Renew Energ 2016;94:579–86.
[42] Zhang J, Cui F, Xu L, Pan X, Wang X, Zhang X, Cui T. Nanoscale 2017;26:15990–7.
[43] Guo PM, Huang FH, Zheng MM, Li WL, Huang QD. J Am Oil *Chem Soc* 2012;89:925–33.
[44] Chiang YD, Dutta S, Chen CT, Huang YT, Lin KS, Wu JC, Suzuki N, Yamauchi Y, Wu KC. ChemSusChem 2015;8:789–94.
[45] Wigley TML, Jones PD, Kelly PM. Nature 1981;291:285.
[46] Gardy J, Hassanpour A, Lai XJ, Ahmed MH. Appl *Catal* A 2016;527:81–95.
[47] Gardy J, Hassanpour A, Lai XJ, Ahmed MH, Rehan M. Appl Catal B 2017;207:297–310.
[48] Madhuvilakku R, Piraman S. Bioresour Technol 2013;150:55–9.
[49] Ali A, Khullar P, Kumar D. Energy sources. Part *A,* recovery. Util Environ *Effects* 2014;36:1999–2008.
[50] Gurunathan B, Ravi A. Bioresour Technol 2015;190:424–8.
[51] Gurunathan B, Ravi A. Bioresour Technol 2015;188:124–7.
[52] Alhassan FH, Rashid U, Taufiq-Yap YH. J Oleo Sci 2015;64:505–14.
[53] Chino M, Nakayama H, Nagai H, Terada H, Katata G, Yamazawa H. J Nucl Sci Technol 2011;48:1129–34.
[54] Labidi S, Ben Amar M, Passarello JP, Le Neindre B, Kanaev A. Ind Eng Chem Res 2017;56:1394–403.
[55] Chen GC, Guo CY, Qiao HB, Ye MF, Qiu XN, Yue CB. Catal Commun 2013;41:70–4.
[56] Saxena SK, Viswanadham N. *Fuel Process* Technol 2014;119:158–65.
[57] Zhou Q, Zhang H, Chang F, Li H, Pan H, Xue W, Hu DY, Yang S. J *Ind Eng Chem* 2015;31:385–92.
[58] Tan T, Lu J, Nie K, Deng L, Wang F. Biotechnol Adv 2010;28:628–34.
[59] Mahmood T, Hussain ST. Afr J Biotechnol 2010;9:858–68.
[60] Kim YK, Lee H. Bioresour Technol 2016;204:139–44.
[61] Wu L, Kim GM, Nishi H, Tatsuma T. Langmuir 2017;33:8976–81.
[62] Chen Z, Wei C, Li S, Diao C, Li W, Kong W, Zhang Z, Zhang W. Nanoscale Res Lett 2016;11:295–302.
[63] Nam SH, Ju DW, Boo JH. J Nanosci Nanotechnol 2014;14:9406–10.
[64] Lei BX, Luo QP, Yu XY, Wu WQ, Su CY, Kuang DB. Phys Chem Chem Phys 2012;14:13175–9.

[65] Lee DY, Shin CY, Yoon SJ, Lee HY, Lee W, Shrestha NK, Lee JK, Han SH. Sci Rep 2014;4:3930–4.

[66] Han GS, Song YH, Jin YU, Lee JW, Park NG, Kang BK, Lee JK, Cho IS, Yoon DH, Jung HS. ACS Appl Mater Interfaces 2015;7:23521–6.

[67] Jang JK, Park SH, Kim C, Ko J, Seo WS, Song H, Park JT. Nanotechnology 2011;22:275720.

[68] Roy AB, Das S, Kundu A, Banerjee C, Mukherjee N. Phys Chem Chem Phys 2017;19:12838–44.

[69] Huang CY, Wang DY, Wang CH, Chen YT, Wang YT, Jiang YT, Yang YJ, Chen CC, Chen YF. ACS Nano 2010;4:5849–54.

[70] Banik A, Ansari MS, Sahu TK, Qureshi M. Phys Chem Chem Phys 2016;18:27818–28.

[71] Cao Y, Denny MS, Caspar JV, Farneth WE, Guo Q, Ionkin AS, Johnson LK, Lu M, Malajovich I, Radu D, Rosenfeld HD, Choudhury KR, Wu W. J Am Chem Soc 2012;134:15644–7.

[72] Yuhas BD, Yang P. J Am Chem Soc 2009;131:3756–61.

[73] Peng Q, Pei K, Han B, Li R, Zhou G, Liu JM, Kempa K, Gao J. Nanoscale Res Lett 2016;11:312–20.

[74] Ossicini S, Amato M, Guerra R, Palummo M, Pulci O. Nanoscale Res Lett 2010;5:1637–49.

[75] Zhang W, Saliba M, Stranks SD, Sun Y, Shi X, Wiesner U, Snaith HJ. Nano Lett 2013;13:4505–10.

[76] Kirkeminde A, Scott R, Ren S. Nanoscale 2012;4:7649–54.

[77] Liu Z, Li J, Sun ZH, Tai G, Lau SP, Yan F. ACS Nano 2012;6:810–8.

[78] Ma W, Luther JM, Zheng H, Wu Y, Alivisatos AP. Nano Lett 2009;9:1699–703.

[79] Du P, Lim JH, Leem JW, Cha SM, Yu JS. Nanoscale Res Lett 2015;10:1030–5.

[80] Liu X, Li J, Zhang Y, Huang J. Chemistry 2015;21:7345–9.

[81] Su R, Tiruvalam R, Logsdail AJ, He Q, Downing CA, Jensen MT, Dimitratos N, Kesavan L, Wells PP, Bechstein R, Jensen HH, Wendt S, Catlow CR, Kiely CJ, Hutchings GJ, Besenbacher F. ACS Nano 2014;8:3490–7.

[82] Wu M-C, Hiltunen J, Sápi A, Avila A, Larsson W, Liao H-C, Huuhtanen M, Tóth G, Shchukarev A, Laufer N, Kukovecz A, Kónya Z, Mikkola J-P, Keiski R, Su W-F, Chen Y-F, Jantunen H, Ajayan PM, Vajtai R, Kordás K. ACS Nano 2011;5:5025–30.

[83] Liu X, Liu Z, Lu J, Wu X, Chu W. J Colloid Interface Sci 2014;413:17–23.

[84] deKrafft KE, Wang C, Lin W. Adv Mater 2012;24:2014–8.

[85] Long L, Yu X, Wu L. Li J, Li X. Nanotechnology 2014;25: 035603.

[86] Jovic V, Smith KE, Idriss H, Waterhouse GI. ChemSusChem 2015;8:2551–9.

[87] Rico-Oller B, Boudjemaa A, Bahruji H, Kebir M, Prashar S, Bachari K, Fajardo M, Gómez-Ruiz S. Sci Total Environ 2016;563:921–32.

[88] Li Q, Meng H, Yu J, Xiao W, Zheng Y, Wang J. Chemistry 2014;20:1176–85.

[89] Wang F, Zheng M, Zhu C, Zhang B, Chen W, Ma L, Shen W. Nanotechnology 2015;26:345402.

[90] Zhang J, Wang Y, Zhang J, Lin Z, Huang F, Yu J. ACS Appl Mater Interfaces 2013;5:1031–7.

[91] Tang X, Chen W, Zu Z, Zang Z, Deng M, Zhu T, Sun K, Sun L, Xue J. Nanoscale 2015;7:18498–503.

[92] Luo M, Lu P, Yao W, Huang C, Xu Q, Wu Q, Kuwahara Y, Yamashita H. ACS Appl Mater Interfaces 2016;8:20667–74.

[93] Mangrulkar PA, Polshettiwar V, Labhsetwar NK, Varma RS, Rayalu SS. Nanoscale 2012;4:5202–9.

[94] Wang Q, Lian J, Li J, Wang R, Huang H. Su B, Lei Z. Sci Rep 2015;5:13593.
[95] Šljukić B, Santos DM, Vujković M, Amaral L, Rocha RP, Sequeira CA, Figueiredo JL. ChemSusChem 2016;9:1200–8.
[96] Shi J, Kuwahara Y, Wen M, Navlani-García M, Mori K, An T, Yamashita H. Chem Asian J 2016;11:2377–81.
[97] Gao S, Li GD, Liu Y, Chen H, Feng LL, Wang Y, Yang M, Wang D, Wang S, Zou X. Nanoscale 2015;7:2306–16.
[98] Abo-Hamed EK, Pennycook T, Vaynzof Y, Toprakcioglu C, Koutsioubas A, Scherman OA. Small 2014;10:3145–52.
[99] Kandiel TA, Anjum DH, Takanabe K. ChemSusChem 2014;7:3112–21.
[100] Erogbogbo F, Lin T, Tucciarone PM, LaJoie KM, Lai L, Patki GD, Prasad PN, Swihart MT. Nano Lett 2013;13:451–6.
[101] Reisner E, Powell DJ, Cavazza C, Fontecilla-Camps JC, Armstrong FA. J Am Chem Soc 2009;131:18457–66.
[102] Caputo CA, Wang L, Beranek R, Reisner E. Chem Sci 2015;6:5690–4.
[103] Cuendet P, Rao KK, Grätzel M, Hall DO. Biochimie 1986;68:217–21.
[104] Brown KA, Dayal S, Ai X, Rumbles G, King PW. J Am Chem Soc 2010;132:9672–80.
[105] Brown KA, Wilker MB, Boehm M, Dukovic G, King PW. J Am Chem Soc 2012;134:5627–36.
[106] Caputo CA, Gross MA, Lau VW, Cavazza C, Lotsch BV, Reisner E. Angew Chem Weinheim Bergstr Ger 2014;126:11722–6.
[107] Baker SE, Hopkins RC, Blanchette CD, Walsworth VL, Sumbad R, Fischer NO, Kuhn EA, Coleman M, Chromy BA, Létant SE, Hoeprich PD, Adams MW, Henderson PT. J Am Chem Soc 2009;131:7508–9.
[108] Mittelbach M, Remschmidt C. Biodiesel - the comprehensive handbook (1st Ed.) Börsedruk Ges. M.b.H; Vienna. 2004.
[109] Wang CM, Yeh KL, Tsai SJ, Jhan YL, Chou CH. Molecules 2017;22:1–13.
[110] El-Batal AI, Farrag AA, Elsayed MA. El-Khawaga AM Bioengineering (Basel) 2016;12:1–24.
[111] Mukherjee J, Gupta MN. Bioresour Technol 2016;209:166–71.
[112] Jiang Y, Sun W, Zhou L, Ma L, He Y, Gao J. Appl Biochem Biotechnol 2016;179:1155–69.
[113] Pang J, Zhou G, Liu R, Li T. Korean J Couns Psychother 2016;59:35–42.
[114] Bencze LC, Bartha-Vári JH, Katona G, Toşa MI, Paizs C, Irimie FD. Bioresour Technol 2016;200:853–60.
[115] Rastian Z, Khodadadi AA, Guo Z, Vahabzadeh F, Mortazavi Y. Appl Biochem Biotechnol 2016;178:974–89.
[116] Rastian Z, Khodadadi AA, Guo Z, Vahabzadeh F, Mortazavi Y. PLoS One 2014;9:e115202.
[117] Karimi M, Keyhani A, Akram A, Rahman M, Jenkins B, Stroeve P. Environ Technol 2013;13:2201–11.
[118] Yu CY, Huang LY, Kuan IC, Lee SL. Int J Mol Sci 2013;14:24074–86.
[119] Ngo TP, Li A, Tiew KW, Li Z. Bioresour Technol 2013;145:233–9.
[120] Tran DT, Chen CL, Chang JS. J Biotechnol 2012;158:112–9.
[121] Dumri K, Hung Anh D. Enzyme Res 2014;(2014)389739.
[122] Sánchez-Ramírez J, Martínez-Hernández JL, Segura-Ceniceros P, López G, Saade H, Medina-Morales MA, Ramos-González R, Aguilar CN, Ilyina A. Bioprocess Biosyst Eng 2017;40:9–22.
[123] Li Y, Wang XY, Zhang RZ, Zhang XY, Liu W, Xu XM, Zhang YW. J Nanosci Nanotechnol 2014;14:2931–6.

[124] Xu J, Sheng Z, Wang X, Liu X, Xia J, Xiong P, He B. Bioresour Technol 2016;200:1060–4.
[125] Bhattacharya A, Pletschke BI. Enzyme Microb Technol 2014;61:17–27.
[126] Roth HC, Schwaminger SP, Peng F, Berensmeier S. Chemistry Open 2016;5:183–7.
[127] Lima JS, Araújo PH, Sayer C, Souza AA, Viegas AC, de Oliveira D. Bioprocess Biosyst Eng 2017;40:511–8.
[128] Dutta N, Biswas S, Saha MK. Enzyme Microb Technol 2016;95:248–58.
[129] Ahmad R, Sardar M. Indian J Biochem Biophys 2014;51:314–20.
[130] Cabrera MP, Assis CRD, Neri DFM, Pereira CF, Soria F, Carvalho Jr. LB. Biotechnol Rep (Amst) 2017;14:38–46.
[131] Bayramoglu G, Doz T, Ozalp VC, Arica MY. Food Chem 2017;221:1442–50.
[132] Waifalkar PP, Parit SB, Chougale AD, Sahoo SC, Patil PS, Patil PB. J Colloid Interface Sci 2016;482:159–64.
[133] Uzun K, Çevik E, Şenel M, Baykal A. Bioprocess Biosyst Eng 2013;36:1807–16.
[134] Valerio SG, Alves JS, Klein MP, Rodrigues RC, Hertz PF. Carbohydr Polym 2013;92:462–8.
[135] Agrawal R, Srivastava A, Verma AK. J Food Sci Technol 2016;53:3002–12.
[136] Chen T, Yang W, Guo Y, Yuan R, Xu L, Yan Y. Enzyme Microb Technol 2014;63:50–7.
[137] Zhou Y, Yuan S, Liu Q, Yan D, Wang Y, Gao L, Han J, Shi H. Sci Rep 2017;7:1–11.
[138] Verma ML, Chaudhary R, Tsuzuki T, Barrow CJ, Puri M. Bioresour Technol 2013;135:2–6.
[139] Singh RK, Zhang YW, Nguyen NP, Jeya M, Lee JK. Appl Microbiol Biotechnol 2011;89:337–44.
[140] Goh WJ, Makam VS, Hu J, Kang L, Zheng M, Yoong SL, Udalagama CN, Pastorin G. Langmuir 2012;28:16864–73.
[141] Dhiman SS, Kalyani D, Jagtap SS, Haw JR, Kang YC, Lee JK. Appl Microbiol Biotechnol 2013;97:1081–91.
[142] Landarani-Isfahani A, Taheri-Kafrani A, Amini M, Mirkhani V, Moghadam M, Soozanipour A, Razmjou A. Langmuir 2015;31:9219–27.
[143] Dhiman SS, Jagtap SS, Jeya M, Haw JR, Kang YC, Lee JK. Biotechnol Lett 2012;34:1307–13.
[144] Rafindadi AA, Ozturk I. Renew Sustain Energy Rev 2017;75:1130–41.
[145] Lee M, Hong T, Yoo H, Koo C, Kim J, Jeong K, Jeong J, Ji C. Renew Sustain Energy Rev 2017;75:1066–80.
[146] Tanatvanit S, Limmeechokchai B, Chungpaibulpatana S. Renew Sustain Energy Rev 2003;7:367–95.
[147] Bugaje IM. Renew Sustain Energy Rev 2006;10:603–12.
[148] Sakah M, Diawuo FA, Katzenbach R, Gyamfi S. Renew Sustain Energy Rev 2017;79:544–57.
[149] Sinha A. Renew Sustain Energy Rev 2017;79:9–14.
[150] Bhatia SK, Bhatia RK, Yang YH. Renew Sustain Energy Rev 2017;79:1078–90.
[151] Hu Y, Cheng H. Nat Commun 2017;8:14590.

CHAPTER

Wastewater remediation via combo-technology

3

Alka Dwevedi*,†, Arvind M. Kayastha‡

*Department of Biotechnology, Delhi Technological University, New Delhi, India**

Swami Shraddhanand College, University of Delhi, New Delhi, India†

School of Biotechnology, Institute of Science, Banaras Hindu University, Varanasi, India‡

3.1 Introduction

The availability of safe water has been an ongoing problem and is becoming worse with increasing urbanization and population density. Only about 30% of the fresh water total is available on earth, in groundwater and surface water, for drinking and other daily human activities as well as industrial and agricultural activities. However, only 2% of fresh water is available for usage due to its nonavailability (comes at inconvenient times and places) and contamination by human activities (Fig. 3.1). Developed countries have advanced technology for water treatment and are thus able to manage the problem of water scarcity to a large extent, but advanced technology is largely unprocurable in developing and undeveloped countries due to cost. Based on the current population growth rate, it has been estimated that over 3.5 billion people will be in a water scarcity condition by 2025. The problem has become exacerbated due to the introduction of various recalcitrant, nondegradable compounds by agricultural and industrial activities. These compounds cannot be removed completely by any current available technologies.

Contaminated water has a high concentration of heavy metals, phenolic compounds, pesticides, etc., in addition to a high microbial load. Children (0–8 years) are at major risk from consumption of contaminated water leading to various neurological diseases, weakening of the immune system, and arrested growth [1] (Table 3.1). Over 1.8 million (4.1% of total global deaths due to diseases) human deaths have been reported by WHO (World Health Organization) annually due to consumption of contaminated water. The number of deaths due to waterborne diseases is directly correlated with socioeconomic condition: the lower the socioeconomic condition, the higher the number of deaths due to waterborne diseases [2]. The available conventional water treatment technologies, like solvent extraction, activated carbon adsorption, chemical oxidation, or biological methods, are inadequate to effectively remove recalcitrant pollutants from water.

Solutions to Environmental Problems Involving Nanotechnology and Enzyme Technology
https://doi.org/10.1016/B978-0-12-813123-7.00003-7

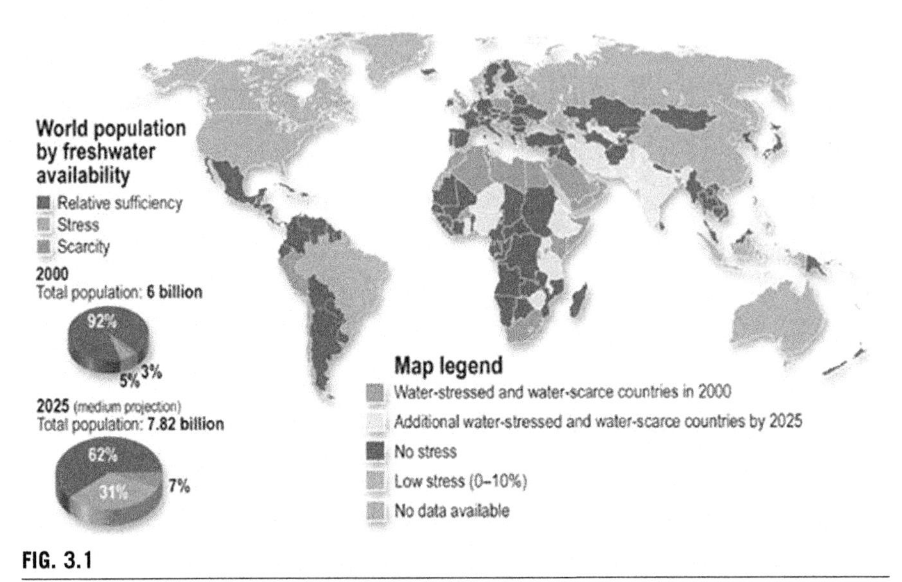

FIG. 3.1

An overview of availability of fresh water in the world.

Adapted from https://www.nanowerk.com/spotlight/spotid=4662.php.

Table 3.1 Summary of Diseases Caused by Consumption of Polluted Water

Diseases	Microbial Agent	Symptoms
Amoebiasis	*Schistosoma* sp.	Abdominal discomfort, fatigue, weight loss, diarrhea, bloating, fever
Cryptosporidiosis	*Dracunculus medinensis*	Flu-like symptoms, watery diarrhea, loss of appetite, weight loss, bloating, nausea
Cyclosporiasis	*Taenia* sp.	Cramps, nausea, vomiting, muscle aches, fever, and fatigue
Giardiasis	*Fasciolopsis buski*	Diarrhea, abdominal discomfort, bloating, and flatulence
Microsporidiosis	*Hymenolepis nana*	Diarrhea and wasting in immunocompromised individuals
Schistosomiasis	*Echinococcus granulosus*	Blood in urine, fever, chills, cough, itchy skin
Dracunculiasis	*Ascaris lumbricoides*	Allergy, rash, nausea, vomiting, diarrhea, asthmatic attack
Taeniasis	*Enterobus vermicularis*	Intestinal disturbances, neurological manifestations, loss of weight
Fasciolopsiasis	*Fasciolopsis buski*	Diarrhea, liver enlargement, obstructive jaundice, cholecystitis
Hymenolepiasis	*Hymenolepis nana*	Abdominal pain, severe weight loss, itching in the anus

Table 3.1 Summary of Diseases Caused by Consumption of Polluted Water—cont'd

Diseases	Microbial Agent	Symptoms
Echinococcosis	*Echinococcus granulosus*	Liver enlargement, anaphylactic shock
Ascariasis	*Ascaris lumbricoides*	Fever, diarrhea, vomiting, underdevelopment
Enterobiasis	*Enterobius vermicularis*	Perianal itch, nervous irritability, hyperactivity, insomnia
Botulism	*Clostridium botulinum*	Dry mouth, blurred vision, difficulty in swallowing, muscle weakness, diarrhea, vomiting
Campylobacteriosis	*Campylobacter jejuni*	Dysentery-like symptoms, high fever
Cholera	*Vibrio cholerae*	Watery diarrhea, nausea, cramps, nosebleed, vomiting, hypovolemic shock, death
E. coli infection	*E. coli*	Diarrhea, death in immunocompromised individuals
M. marinum infection	*Mycobacterium marinum*	Lesions on elbow, knee, hands, and feet
Dysentery	*Shigella dysenteriae*, *Salmonella* sp.	Blood in the feces, vomiting of blood
Legionellosis	*Legionella pneumophila*	Resembles acute influenza without pneumonia, muscle ache, diarrhea
Leptospirosis	*Leptospira*	Flu-like symptoms followed by meningitis, liver damage, renal failure
Otitis externa	*Number of bacterial and fungal species*	Ear canal swells, causing pain and tenderness
Salmonellosis	*Salmonella* sp.	Diarrhea, fever, vomiting, abdominal cramps
Typhoid fever	*Salmonella typhi*	Fever, sweating, diarrhea, sometimes death
Vibrio illness	*Vibrio vulnificus, V. alginolyticus, V. parahaemolyticus*	Tenderness, agitation, bloody stools, chills, hallucination, fatigue, weakness
Severe acute respiratory syndrome (SARS)	*Coronavirus*	Fever, myalgia, lethargy, cough, sore throat
Hepatitis	*Hepatitis A virus*	Fatigue, fever, nausea, diarrhea, weight loss, itching, jaundice
Poliomyelitis	*Poliovirus*	Headache, fever, spastic paralysis, sometimes death
Desmodesmus infection	*Desmodesmus armatus*	Fever, diarrhea, vomiting, weakness

Nanotechnology can provide a solution to these water treatment issues by removal of many types of recalcitrant compounds in addition to microbial load, including viruses. Besides water treatments using nanomaterials, such as carbon nanotubes (CNTs), nanosorbents, dendrimers, etc., nanotechnology can also be used in water desalination, disinfection, and sensors (contaminants can be sensed even at sub-ppm concentrations). Four classes of nanomaterials are now being used for water treatment: metal-containing nanoparticles, carbonaceous nanomaterials, zeolites, and dendrimers. They can remove even toxic heavy metals like arsenic, organic material, salinity, nitrates, pesticides, etc. from surface water, groundwater, and wastewater. This is true because of their large specific surface area, high reactivity, high degree of functionalization, size-dependent properties, and high affinity for specific target contaminants. Most importantly, nanomaterials can be used to treat water without addition of chlorine (known to be controversial due to generation of carcinogenic compounds). The technology of water desalination, softening, and salt recovery as well as treatment of industrial and brackish water using nanostructured filters and nanoreactive membranes have made nanotechnology more in demand [3]. Electrospun nanofibers have gained much attention in water treatment, as they also can inhibit the bacterial growth responsible for membrane biofouling. Nanomaterials used for water treatment can easily be merged into distributed optimal technology networks (DOTNETs) being used in water treatment for small-scale housing subdivisions, apartment complexes, and commercial districts. Further, they can also be used in treatment plants like factory assembled, compact ready-to-use water treatment systems, point of entry (POE), and point of use (POU) treatment units.

The potential significance of nanotechnology in water treatment is huge, with several hundreds of publications and registered patents already. However, this technology has not yet reached the commercial market. The foremost reason for this is the potential toxicological risks for humans and the environment, which cannot be ignored for the sake of the many technological advantages. Genotoxicity and cytotoxicity studies have revealed that nanomaterials are effective agents in interacting with biological macromolecules present in living systems and thus are responsible for a number of diseases and clinical disorders [4]. Therefore a need exists for additional technology that can lower the toxicological impact of nanomaterials while retaining their abilities for water purification, disinfection, sensing, and monitoring.

An additional issue related to water purification is removal of biofilm (extracellular polymeric substance secreted by sessile bacterial communities) developed over a course of time. The development of biofilm is responsible for decreasing the efficiency of membranes derived from nanomaterials, with an additional risk of adding metabolic products and biological toxins that degrade water quality [5]. Biofilm is most frequently formed in water coming from the medical, paper, and food-processing industries. The available antimicrobials or chemicals are not efficacious in combating biofilm. Even mechanical methods have not helped in this respect, and they make the process costly and more labor intensive.

Enzyme technology, based on suitable enzymes being immobilized onto nanomaterials, can help in removing the toxicity of nanomaterials, as well as providing efficacy in removal of biofilm by degrading its components containing polysaccharides and proteins. Enzyme technology has been found to complement nanotechnology by enhancing water-purifying capabilities by several-thousandfold, due to its excellent specificity and catalytic efficiency. Enzymes have further benefited this combination by increasing the stability and number of cycles of reusability. This chapter outlines water purification, disinfection, sensing, and monitoring using the combination of the two technologies, nanotechnology and enzyme technology.

3.2 Overview on wastewater remediation across the world

Wastewater is the used water discharged from various sources including homes, businesses, industries, cities, and agriculture. There are various types of wastewater based on water uses: urban wastewater (blackwater coming from toilets and greywater coming from kitchens and bathing), industrial wastewater, commercial establishment wastewater (including hospitals), agricultural wastewater, etc. Further, collected wastewater in a municipal piped system (sewerage) is called sewage. Wastewater treatment usually follows three steps: primary treatment, secondary treatment, and tertiary treatment. Before proceeding through the various steps, wastewater is pretreated to remove grit, grease, and gross solids that could hinder the subsequent treatment stages. *Primary treatment* aims to settle down suspended solids (both organic and inorganic), followed by removing them. Primary settlers, septic tanks, and imhoff tanks are most commonly used for primary water treatment. *Secondary treatment* involves removal of soluble biodegradable organics through degradation by bacteria and protozoa, through aerobic or anaerobic biological processes. Aerated lagoons, activated sludge, trickling filters, oxidation ditches, etc. are used for secondary water treatment. *Tertiary treatment* polishes the effluent before it is discharged or reused, by removal of nutrients (mainly nitrogen and phosphorous), toxic compounds, residual suspended matter, or microorganisms, through disinfection with chlorine, ozone, ultraviolet radiation, or membrane filtration (micro-, nano-, ultra-, and reverse osmosis), etc. Tertiary treatment is the most costly of the three treatments, and is often not done in low-income countries. Finally, the treated water is released for various applications.

The need to provide accurate data on levels of wastewater generated annually has been emphasized by national policymakers, researchers, practitioners, and public institutions, as this data is necessary for developing national policy and action plans based on type of wastewater and contaminant levels. However, accurate information related to wastewater is not systemically monitored in most of the countries of the world, with significant data paucity in rural areas. In 2010, global annual domestic water withdrawals modeled by WaterGAP3 (global freshwater model that calculates flows and storages of water across the world) have found that there was approximately 450 km^3 of global production of wastewater coming from the domestic and manufacturing sectors, with 70% of wastewater (315 km^3) coming from the domestic sector. AQUASTAT (database that provides information on global water systems) has reported that globally more than 330 $km^3\ year^{-1}$ of (mostly) municipal wastewater are produced [6]. Countries like the United States, China, Brazil, Japan, Russia, India, and Indonesia collectively produce over 167 km^3 of wastewater per year (half of global municipal wastewater production). It has also been reported by AQUASTAT that, on average, 60% of produced municipal wastewater is treated globally. These figures on wastewater treatment are much lower due to the poor treatment capacities of wastewater treatment plants, particularly in middle- and low-income countries. Further, data is generally not reported by some low-income countries with large urban populations (such as Nigeria). It has been found that most of the countries report only secondary and tertiary treated wastewater as "treated wastewater," while some countries report only primary treated wastewater, which makes data aggregation and comparisons by country rather more difficult. In 2014, GWS (global water store) 2009 reported that globally 24 $km^3\ year^{-1}$ of wastewater generally passes all three phases of water treatment [7]. It has been found from wastewater treatment data that there is a strong correlation between treatments of wastewater and a country's income [8]. High-income, middle-income, and low-income countries have reported an average of wastewater treatment as 70%, 28%, and 8%, respectively. More importantly, in the case of developing countries, most untreated wastewater is directly discharged into natural water bodies due to either dysfunctionality or limited capacity of water treatment plants.

Increasing water scarcity has been a factor in the growing realization of the necessity of wastewater treatment. However, the efficiency of wastewater treatment is directly related to physical, economic, social, regulatory, and political environments. It has been found that high-income countries follow complete protocols for wastewater treatment and the process is well regulated and planned (Fig. 3.2). The usage of wastewater is based on the economic status of the countries, as discussed in the following paragraphs.

- *Direct use of untreated wastewater*: This occurs in water-scarce areas having drier climates, usually in low-income countries. Lack of water sources, groundwater salinity, and unaffordability of fresh water are the major reasons behind usage of untreated wastewater. Most commonly, untreated wastewater is used for agriculture [9]. There are various examples worldwide, particularly in developing and undeveloped countries. A number of farms in Pakistan, for example, are using untreated wastewater due to the extreme salinity of the groundwater; farmers of the twin cities of Hubli-Dharwad (Karnataka, India), living in a semiarid climate, use untreated wastewater coming from open sewers and underground sewers for irrigation. Similarly, farmers from Cochabamba in Bolivia and Accra and Tamale in Ghana use wastewater from malfunctioning treatment plants or sewers. It has been found that in Haroonabad (Pakistan) and Hyderabad (India), wastewater is the only water flowing in the irrigation canals during the dry season. Even worse, in Nairobi (Kenya), Maili Saba (Kenya), Addis Ababa (Ethiopia), Bhaktapur (Katmandu Valley in Nepal), and Dakar (Senegal), farmers have removed sewage line inspection covers that block sewers, followed by the rising of raw sewage up to the manholes, which subsequently flows into their farmland [10].

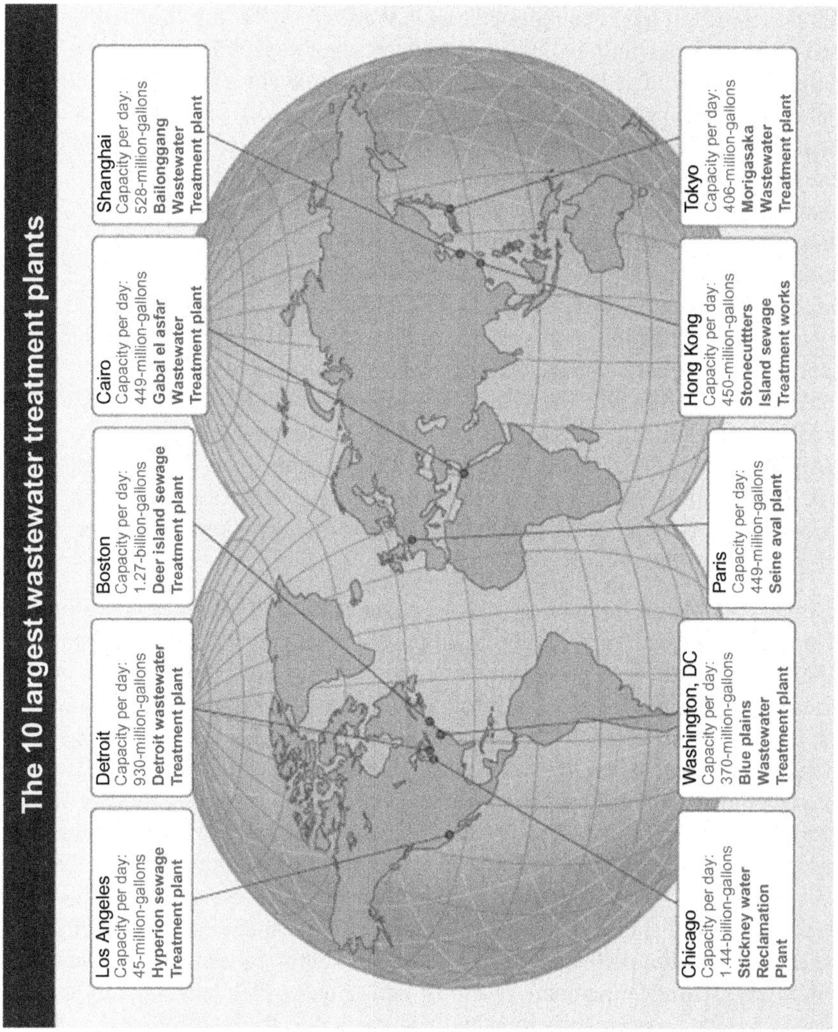

FIG. 3.2

The 10 largest wastewater treatment plants in the world.

Reprinted courtesy of Engineering News Record, Copyright BNP Media, April 2, 2012, All rights reserved.

- *Indirect use of untreated wastewater*: This is being practiced in low- and middle-income countries having drier and wetter climates. Here, untreated wastewater is directly discharged into freshwater streams for dilution and is subsequently used by farmers, in households, or by industries. Several examples of indirect use of untreated wastewater are found in sub-Saharan Africa, Nepal, India, and many cities in Brazil, Argentina, and Colombia that have inadequate sanitation facilities. It has been reported that in West Africa, farmers use highly polluted water in the vicinity of cities for irrigating their vegetables. It has been found that up to 90% of vegetables being consumed in the cities are grown through irrigation using polluted water [11].
- *Planned use of reclaimed water*: This occurs more frequently in higher income countries having well-established water treatment plants. They are much more environmentally aware of their wastewater, which is reclaimed and used to preserve freshwater ecosystems. Reclaimed water is directly used for agricultural irrigation, city landscaping, golf courses, toilet flushing, washing of vehicles, and groundwater recharge, as well as being a source of potable water supply. Windhoek in Namibia is one of the examples of this type of water reclamation and usage. Even industrial wastewater is purified to industrial standards and recycled within the system. In countries of the Middle East and North Africa, Australia, the Mediterranean, and the United States of America (AQUASTAT 2014; Global Water Intelligence 2010), planned use of reclaimed water is seen. In all of these cases, an effective sanitation and treatment technology supports water reclamation; however, the main challenge for water reuse is public acceptance [12].

Water treatment plants, besides producing clean water, also produce lots of sludge, which can be used for a variety of purposes. Even excreta collected from toilets either on-site (e.g., in a pit latrine or septic tank) or transported off-site in sewer systems are also termed sludge. Fresh and untreated sludge contains many pathogens, has a high proportion of water, has higher BOD (biochemical, or biological, oxygen demand), and is generally putrid and odorous. Additionally, it also contains essential nutrients like nitrogen and phosphorous, so it is a very beneficial fertilizer for plants. The organic carbon present in sludge is used as a soil conditioner, as it improves soil structure that helps in proper growth of plant roots. Further, the organic carbon can also be converted into energy through biodigestion or incineration. The composition of sludge depends on the source types; sometimes it is highly contaminated with harmful pollutants like heavy metals (in the case of sewage coming from industries). Sludge treatment is a must before using the sludge in various applications. The treatment involves reducing water content, BOD, pathogens, and obnoxious odor, followed by thickening, drying, and stabilization. Dewatering is an energy consumptive process, sometimes taking even weeks for complete sludge drying through incineration, pyrolysis, or gasification [13]. The most important step, responsible for making sludge fit for various applications, is stabilization, accomplished by aerobic and anaerobic processes. Aerobic stabilization uses composting at higher temperatures (55°C), similar to the natural process that takes place in the forest leading to breakdown of organic material containing leaf litter, animal wastes, etc. and reducing volume by severalfold. In the case of anaerobic stabilization (cannot be used when contaminated with heavy metals), reduction of organic wastes takes place through bacterial decomposition under anaerobic conditions and produces a mixture of gases containing methane, carbon dioxide gas, and syngas. Properly treated sludge is called biosolids. These solids have a variety of beneficial uses like landscaping, energy recovery, generation of new soil, increasing soil fertility, etc. Sludge treatment is also dependent on the economic status of the country [14].

- *Informal use of untreated sludge*: Many low-income countries in West Africa and South Asia do follow the complete steps for sludge treatment. However, in these countries sludge is collected and air dried and directly used as fertilizer in the fields. In some cases, sludge fills vacant land, leading to anaerobic degradation by natural processes in an unplanned way. In other cases, sludge is discharged directly into water bodies, where it is diluted and finds its way back into the food chain when that water is used for farming.
- *Formal use of biosolids*: This occurs in developed countries with regulated and well-designed biosolid and septage programs. Properly planned sludge treatment leads to the generation of biosolids, which is very useful in producing soil nutrients and thus maintains soil fertility while reducing the pressure on the final disposal sites. Energy recovery from sludge through biodigestion and incineration has an additional advantage over reuse of sludge. However, there is the risk of contamination by heavy metals and toxic pollutants, so thorough studies on composition of the sludge must be done before sludge processing occurs.

3.3 Nanotechnology and water remediation

Nanomaterials (size $<100\,nm$) are found to be effective in water treatment due to their high specific surface area, high reactivity, and excellent sorption [15]. Zeolites, CNTs, biopolymers, self-assembled monolayer on mesoporous supports, zero-valent iron nanoparticles, bimetallic iron nanoparticles, and nanoscale semiconductor photocatalysts are the most commonly used nanoparticles in water treatment [16] (Tables 3.2 and 3.3). Further, nanomaterial capabilities in desalination and their easier incorporation into existing technology has made them effective. Even industrial and brackish water can be made potable in a very short time span with little labor in a cost-effective way. Water treatment using membranes based on nanofibers have been widely accepted due to their efficacy and easier regeneration. However, the accuracy of the number of recycles using regeneration protocols has to be closely monitored, in order to nullify the probability of nanomaterials leaching into the treated water after a certain time. The following paragraphs provide brief details on water treatment using nanomaterials.

Table 3.2 Summary of Nanomaterial Applications in Wastewater Treatment

Applications	Nanomaterials	Enabled Technologies
Adsorption	Carbon nanotubes (CNTs), nanoscale metal oxides, nanofibers with core shell structure	Adsorption of recalcitrant contaminants, for adsorptive media filters and slurry reactors
Membrane processes	Nanozeolites, nano-Ag, CNTs, nano-TiO_2, aquaporin, nanomagnetite	Antibiofouling with stringent permeability capabilities
Photocatalysis	Nano-TiO_2, fullerene derivatives	For photocatalytic reactors, solar disinfection
Disinfection and microbial control	Nano-Ag, CNTs, nano-TiO_2	POU water disinfection, antibiofouling surface
Sensing and monitoring	Quantum dots, CNTs, silica nanoparticles, magnetic nanoparticles, noble metal nanoparticles	Optical and electrochemical detection

Table 3.3 Mechanisms Adopted by Nanomaterials for Effective Waterborne Pathogen Killing

Nanomaterials	Antimicrobial mechanisms
Nano-Ag	Release of silver ions, protein damage, suppress DNA replication, membrane damage
Nano-TiO_2	Production of reactive oxygen species (ROS)
Nano-ZnO	Release of zinc ions, production of H_2O_2, membrane damage
Nano-MgO	Membrane damage
Nano-Ce_2O_4	Membrane damage
Fullerol and aminofullerene	Production of ROS
CNTs	Membrane damage, oxidative stress
Graphene-based nanomaterials	Membrane damage, oxidative stress

(i) Remediation and Desalinization

(a) *Adsorption*: Nanoparticles are excellent adsorbents due to their extremely large specific surface area (size: 10–100 nm), high selectivity, short intraparticle diffusion distance, tunable pore size, and adsorption kinetics. They can remove recalcitrant organic and inorganic contaminants, degrade organic compounds (chlorinated alkanes and benzenes, pesticides, organic dyes, nitro aromatics, dioxins, furans, etc.) and reduce toxic metal ions [Cr (VI), Ag (I), Pt (II)] [17]. Nanoparticles can easily be functionalized with various chemical groups depending on the contaminant types (making them more efficient). Commonly used nanoparticles for adsorbents are: TiO_2, Fe^0, Fe^0/Pd^0, Fe^0/Pt^0, Fe^0/Ag^0, Fe^0/Ni^0, Fe^0/Co^0, or hybrids (*viz.* $Pt/TiO_2/Ru$ nanoparticles). Innovative nanosorbents prepared by hybridization of inorganic and organic nanosorbents (*viz.* sodium dodecyl sulfate incorporated into magnesium-aluminum-layered double hydroxides) are found to have excellent sorption capacity for contaminants like tetrachloroethylene (PCE) and trichloroethylene (TCE) [18]. CNTs are used for sorption of metal ions like Pb (II), Cu (II), and Cd (II) with several thousand higher sorption capacities than the powdered or granular activated carbon. Further, they can adsorb volatile organic compounds and water-soluble dyes (acridine orange, ethidium bromide, eosin bluish, and orange G). Chitosan nanoparticles (40–100 nm) can easily be functionalized based on the application type, *viz.* phosphate functionalization used to adsorb Pb (II) [19]. Nanozeolites (*viz.* NaP1 zeolites: $Na_6Al_6\ Si_{10}O_{32}$, $12H_2O$) are used for removal of heavy metal ions (Cr, Ni, Zn, Cu, Cd), even from acidic wastewater, due to their excellent metal ion exchanging capacities. Nanosorbents are available in various forms (pellets, beads, or porous granules), making them more easily incorporated into existing treatment processes. Further, there are well-established protocols for their regeneration, which makes them more desirable due to the increased reusability and cost reduction by severalfold. Arsen X^{np} (hybrid nanoparticles of FeO and polymers) and ADSORBSIA (beaded TiO_2 nanoparticles with diameter from 0.25 to 1.2 mm) for arsenic removal have been commercialized on a pilot scale [20]. Following are brief details on nanosorbents:

- Carbon-based nanosorbents: The most commonly used carbon-based nanosorbent is the CNT. It is very effective in removal of bulky organic compounds, including antibiotics and various pharmaceutical compounds and heavy metal (Cu^{2+}, Pb^{2+}, Cd^{2+}, and Zn^{2+}) contaminants, more so than conventionally used activated carbon. This is attributed to the very large surface area and the number of interactions, including hydrophobic effect, π-π interactions, hydrogen bonding, covalent bonding, and electrostatic interactions [21]. It has fast adsorption kinetics due to the large number of accessible adsorption sites containing functional groups like carboxyl, hydroxyl, and phenol, as well as the small distance of intraparticle diffusion [22]. However, the frequent problem of CNT aggregation in aqueous media decreases its effective surface area. In terms of adsorption, there is little difference with CNT aggregation due to the presence of many interstitial spaces and grooves helping in adsorption [23]. For many specific applications, CNTs are coated with specific agents like sand granules, which helps in removal of toxic metal ions (Hg^{2+}) and even removal of bulky dyes like rhodamine B [24]. Generally, carbon-based nanosorbents are used during the polishing step of water treatment due to their high cost and sensitivity.
- Metal-based nanosorbents: These are generally used for heavy metals (arsenic, lead, mercury, copper, cadmium, chromium, and nickel) and radionuclide removal from wastewater. Metal oxides (FeO, TiO_2, and Al_2O_3) are commonly used as adsorbents having excellent kinetics

of metal adsorption due to the large surface area, short intraparticle distance of diffusion, and large number of reaction sites [25]. Metal oxide nanosorbents have the additional property of magnetism, giving them more efficiency in removal of heavy metals. It has been found that the smaller the size of nanosorbents, the greater is the magnetism and thus the higher the adsorption capacity. Lab experimentation has found that a decrease in particle size increases the volume of many atomic magnetic dipoles. However, if the size is less than 11 nm, the particle loses its own magnetism but responds to externally applied magnetism [26, 27]. Metal oxide nanosorbents can be easily compressed into porous pellets and fine powders using moderate pressure, based on the type of application. These nanosorbents are recommended for removal of arsenic and organic contaminants for POU applications. Further, their low cost and low toxicity have made them very popular.

- Polymeric nanosorbents: The most commonly used polymeric nanosorbents are dendrimeric nanosorbents for treatment of water containing organic compounds and heavy metals. Their inner shell is hydrophobic, which adsorbs organic compounds, while the hydrophilic exterior shell adsorbs heavy metals through its tailored branches. The phenomenon of adsorption is caused by complexation, electrostatic interactions, hydrophobic effect, and hydrogen bonding. Some dendrimer nanosorbents have been prepared for removal of specific contaminants; for example, polyamidoamine (PAMAM) dendrimer-NH_2 is used for removal of copper ions from wastewater. The main advantage of regeneration is desirable due to the significant cost reduction. Lab testing on dendrimer-based water treatment is in process; hopefully dendrimers will soon make it into the commercial market [28].

(b) *Membranes and membrane processes*: Membranes and their related processes make use of nanomaterials like nano-Ag, CNTs, and polyvinyl-*N*-carbazole-SWNT (single walled nanotubes) nanocomposites. They carry out water treatment by providing a physical barrier using selectivity and permeability. They can remove organic and inorganic pollutants and heavy metals as well as biological contaminants (bacteria, virus) from groundwater and surface water. Membranes have wide credence due to their nontoxicity, reusability, minimal labor requirements for operation and maintenance, small space requirements, and ability to fit easily in any set-up. Further, they are mechanically and thermally stable, resistant to biofouling, and have excellent shelf-life [29]. Water purification and desalination are carried out by these membranes using ultrafiltration (UF), nanofiltration (NF), forward osmosis (FO), and reverse osmosis (RO). Only RO and FO are energy-driven processes, while the others work on the differences in the osmolalities. They have made it possible to even treat sea water and brackish water and make it fit for drinking [30]. Sometimes doping with suitable materials is required to prevent any nanomaterials leaching into the treated water; this also can inactivate microbial pathogens [31]. The following paragraphs discuss different types of membranes and their applications.

- Nanofiber membranes: These are made of electrospinning ultrafine fibers derived from polymers, ceramics, and metals. They are highly porous and have very large surface area. Their physical properties, like diameter, composition, morphology, secondary structure, and spatial alignment, can easily be modulated, due to which they have wide applications. They work by UF, FO, and RO processes to purify water containing heavy metals, organic and inorganic pollutants, microbial pathogens, etc. They are also found to be useful in air filtration by removing all harmful gases. They are sometimes doped with materials like ceramics, TiO_2, to protect themselves from biofouling. They are in great demand due to their adaptability and ease of fabrication [32].
- Nanocomposite membranes: These are known for their increased efficacies with respect to various parameters such as membrane permeability, surface hydrophilicity, mechanical and thermal stability, and reduced fouling. A broad range of nanomaterials is used: hydrophilic metal oxide nanoparticles (Al_2O_3, TiO_2, and zeolite), antimicrobial nanoparticles (nano-Ag and CNTs), and (photo) catalytic nanomaterials (bimetallic nanoparticles, TiO_2). Thin film nanocomposite (TFN) membranes have been developed with enhanced membrane permeability and larger surface charge density (mostly negative) based on nanozeolites, nano-Ag, nano-TiO_2, CNTs, etc. Further, they have antifouling properties and increased membrane hydrophilicity, which helps in salt exclusion. Sometimes they are doped to gain additional properties like photocatalysis, higher versatility, etc. based on application type. Quantum Flux (the commercial name of a nanozeolite-based nanocomposite membrane) is used for desalination of sea water and makes it fit for use. It has been doped with nano-TiO_2 (≤ 5 wt%), which prevents membrane fouling [33, 34]. TFN membranes are mostly desirous for their enhanced water permeability and reduced cost. However, they have not been very successful in water desalination due to various limitations [35].
- Biologically inspired membranes: These are derived from biological systems and have excellent selectivity and permeability. The most commonly used biological membrane is aquaporin-Z derived from *Escherichia coli*. It contains amphiphilic triblock-polymer vesicles with complete impermeability to small molecular weight solutes like glucose, glycerol, salt, and urea. For commercial applications, biological membranes are incorporated with nanomaterials, *viz.* CNTs. Aquaporins with aligned CNTs have increased water permeability with very high selectivity for small molecular weight solutes. They can be used in water desalination and for removal of various molecules. However, their efficacy is directly related to the uniformity of CNTs for reliable

salt rejection from seawater. Further, their chemical modification of CNTs helps in improving solute selectivity. For example, the carboxyl functional group rejects 98% of $Fe(CN)_6^{3-}$, 50% for KCl at 0.3 mM, and negligible when KCl is at 10 mM. There are various technical challenges that have impeded large production of aquaporins with aligned CNTs [36].

- Nanostructured and reactive membranes: These membranes are existing filtration membranes fabricated with various nanomaterials to enhance efficiency. For example, depositing nano-TiO_2 improves contaminant degradation in the presence of sunlight; metal retention capacity can be enhanced by depositing poly (L-glutamic acid) and poly (L-lysine); and fabrication with CNTs helps in removal of pathogenic microorganisms like *E. coli*, *Staphylococus aureus*, and Poliovirus sabin 1 from contaminated water. These membranes are very useful in water treatment for removal of those recalcitrant contaminants that cannot be removed by any other known methods. Alumina membranes fabricated with several layers of poly(styrene sulfonate)/poly(allylamine hydrochloride) have high retention for divalent cations (Ca^{2+} and Mg^{2+}) and anions like Cl^- and SO_4^{2-} while fabrication with A-alumoxanes nanoparticles (7–25 nm) leads to increased selectivity for synthetic dyes like Direct Red 81, Direct Yellow 71, and Direct Blue 71. Incorporation of bimetallic Fe^0/Pt^0 nanoparticles or zero-valent iron (nZVI) into a cellulose acetate membrane helps in reduction of chlorinated organic compounds like TCE (trichloroethylene), PCE (tetrachloroethylene), etc. More innovations are yet to come, with the additional advantage of wide acceptance due to the presence of the basic membranes, which are commonly used [37].

(ii) Photocatalysis for Wastewater Treatment

Photocatalysis (oxidation in the presence of sunlight) helps to kill pathogens using oxidation and degradation of recalcitrant organic compounds as well as hazardous nonbiodegradable contaminants. Nanomaterials like TiO_2, CeO_2 and CNTs have excellent photocatalytic properties due to their large surface area, which can be used for water treatment by completely degrading organic pollutants through fast catalysis. TiO_2 is an excellent photocatalyst nanomaterial, as it rapidly generates an ionic pair upon UV irradiation, which forms reactive oxygen species (ROS) responsible for contaminant degradation and pathogenic killing. It is very cost effective, not very toxic, and widely available. The smaller the size of the nano-TiO_2, the higher is the rate of photocatalysis due to the increased number of reactive facets. However, a size smaller than 11 nm does not further increase photocatalysis, as there is an increase in interfacial charge carrier transfer. Further, TiO_2 nanotubes are more efficient than TiO_2-nanoparticles due to the increased photocatalysis and higher degradation rate of organic compounds as provided by shorter diffusion paths inside the tubular wall. It has been found that additional doping with various novel metals, dye sensitizers, semiconductors, and anions can further increase the rate of photocatalysis by narrowing the bandgap, which allows excitation even by visible light. Crystallographic studies on TiO_2 nanomaterials have revealed that {001} facets have the highest photocatalytic activity due to the small spatial separation between electron and holes as well as the strong adsorption [38]. Nano-TiO_2 has been recommended for water treatment on both small and large scales, using various-sized parabolic collectors. Nanomaterials like WO_3 and fullerene derivatives (aminofullerenes and fullerol) are also photocatalytic. However, they are very costly with respect to nano-TiO_2 but have a much narrower bandgap, which allows excitation by visible light. Nano-WO_3 has been commercialized as the Purifics Photo-Cat system to treat 2 million gallons of water per day containing recalcitrant organic compounds, with a power consumption of approximately 4 kWh m^{-3}.

Therefore, nanomaterials having photocatalytic properties could be an effective alternative for water treatment on an industrial scale. It has been recommended to use immobilized photocatalytic nanoparticles in commonly used reactors like slurry reactors, with optimized configuration and operation parameters (reactor design, reaction selectivity, pH, temperature, optimum light wavelength, and intensity). This has an additional advantage of reusability, which helps in lowering the cost and makes the process more effective and more eco-friendly [39].

(iii) Disinfection and Microbial Control

Conventionally, purified treated water is disinfected with strong oxidants like chlorine and ultraviolet (UV) irradiation before its release for consumption. These oxidants carry out pathogen (bacteria and virus) inactivation by blocking their metabolic activities, inhibiting synthesis of housekeeping gene products and disrupting the cell wall, leading to leaking of all cellular constituents. In the case of UV irradiation, DNA synthesis is additionally blocked, leading to disruption of all cellular activities and subsequently the death of the waterborne pathogens. These oxidants are cheap, efficacious, and easily accessible. However, they are also responsible for generation of toxic by-products (trihalomethanes, haloacetic acid, aldehydes), with some of them being carcinogenic in nature, including nitrosamines, bromates, etc. [40]. This condition becomes worse when high dosages are used for heavily pathogenic-loaded water and exceed the admissible range of general usage. Alternatives are being looked for, but all of them are at stage 1 of the Disinfection Byproduct Rule (1996), of the Safe Drinking Water Act amendment.

Nanomaterials like nano-Ag, nano-ZnO, nano-TiO_2, nano-Ce_2O_4, CNTs, and fullerenes are known for their effective antimicrobial properties and for leaving no by-products. They are recommended for controlling pathogens, biofilm formation, and microbial-induced corrosion present in storage tanks and distribution pipes. Different nanoparticles have varied modes of pathogen killing, with a commonality of a broad antimicrobial spectrum. Nano-Ag works by releasing Ag^+ which binds to thiol groups of cellular proteins and inactivates them, leading to death of the pathogens. Additionally, they prevent DNA replication, which inhibits synthesis of important cellular proteins. It has also been found that they can cause structural damage in the pathogenic cell envelope. Nano-Ag has been commercialized under the trade name MARATHON and Aquapure systems, used in POU treatment, microfilters, etc. Despite the low human toxicity, it is crucial to check the level of Ag^+ in the released water. Factors like size, shape, coating, and crystallographic facets are significantly important in the release kinetics of Ag^+. CNTs carry out pathogenic killing through their conductivity leading to cell membrane perturbation, oxidative stress, and disruption of cellular structures [41]. Further, they remove pathogens by size exclusion activity provided by their layered structure. Generally, multiwalled nanotubes (MWNTs) are the most effective due to their fibrous shape and layered structure, which helps in providing excellent conductivity. Mostly small-sized CNTs are preferred for POU devices, due to their higher toxicity than the larger-sized CNTs. Metallic nanoparticles (MgO) can kill Gram-positive and Gram-negative bacteria as well as their spores (e.g., *Bacillus subtillus*) by disrupting their membrane integrity. Nanoparticles like dendrons, dendrimers, hyperbranched polymers, and dendrigraft polymers obstruct pathogens by inhibiting their entry due to their small size (0.1–1.0 nm) provided by their architecture, containing core, interior branch cells, and terminal branch cell. They usually require pressure between 200 and 700 kPa, which makes the process highly energy consumptive. This can be circumvented by a tangential or cross-flow of water movement during purification.

Nanoparticles have been found to be an excellent substitute for water disinfection with respect to other conventional techniques, due to few or no by-products and their higher efficacies. However, they are a little costlier than other available techniques. Various solutions to this have been found, such as reusability of nanoparticles by coating with certain materials that help in their regeneration, and also avoiding any long-term side effects due to leakage. Further, this will also prevent biofouling caused by accumulation of microbes on their surface. Therefore, the nanoparticles could turn out to be the best substitute for present techniques for water purification and disinfection.

(iv) Sensing and Monitoring

Nanosensors are based on nanoparticles, used for checking water quality after their treatment by various methods. They have excellent sensitivity, selectivity, and photostability, and provide data within microseconds. Even harmful pathogens that remain undetected by conventional methods, like hepatitis A and E, coxsackie viruses, echoviruses, adenoviruses, Norwalk viruses, *Legionella*, *Helicobacter*, *Cryptosporidium*, *Giardia*, etc., can be easily detected in a very short time. This efficiency allows the detection of several thousands of samples in a short time period, which is generally not possible by any other available techniques. Thus, they can be used by municipal bodies for water-quality checks on a large scale. Further, they also ensure that even treated water coming from industrial uses can be used without suspicion.

The sensors contain four units: recognition agent, amplifier, processor, and output. Nanomaterials are important components in the last three units due to their electrochemical, optical, and magnetic properties, while recognition is generally handled using antibodies, aptamers, carbohydrates, antimicrobial peptides, etc. Nanomaterials can be used in the first unit of the sensor due to having a very large specific surface area, which can easily be functionalized by various chemical groups for modulating their affinity towards a variety of water contaminants. For example, quantum dots (QDs) modified with TiO_2 are used to detect polycyclic aromatic hydrocarbons at pM, while QDs modified with CoTe are used for bisphenol A at 10 nM, etc. Commonly used nanomaterials for sensors are magnetic nanoparticles, noble metals, silica nanoparticles, dye-doped nanoparticles, CNTs, and QDs. Metal nanoparticle detection is based on changes in absorption at specific wavelengths, while novel metal nanomaterials detect change in Rayleigh scattering due to their localized surface plasmon resonance (LSPR). In the case of silica nanoparticles, surface modification with organic or inorganic luminescent dyes is used to enhance their sensitivity and protect from photobleaching [42]. CNTs are used in the electrode of the sensor, as they have high conductance, which facilitates faster electron transfer in addition to the amplifier unit of the sensor. However, the preparation of homogenous CNTs is the most challenging aspect. They are used for detecting trace metals and organic pollutants in the water. Dynabead is a commercially used nanosensor to detect pathogens. It is based on a magnetic nanocomposite made of CdSe, which is a semiconductor, having excellent electronic characteristics, that detects based on changes in fluorescent spectra. Nanomaterials as sensors can easily enter the market as they are not placed in direct contact with users. However, aspects such as cost effectiveness, regulatory and public acceptance, water solubility, long-term performance and stability, and testing in real natural and wastewater must be researched thoroughly, to bring about their ultimate success in the commercial market. The previously mentioned annoying aspect of aggregation has impeded the use of nanomaterials in nanosensors, as it impairs sensitivity and reproducibility. Work is in progress to obstruct aggregation by surface functionalization, but currently there is no hard-core protocol claiming to resist aggregation completely.

Nanomaterials have provided solutions to water desalination, removal of recalcitrant toxic contaminants and toxic heavy metals, and development of chlorine-free water treatment. Further, their significance in water quality checks makes them desirable for use in various types of water purifiers. The Freedonia group has released a forecast of nanomaterials demand by 2020 by classifying nanomaterials into metal oxides, clays, metals, polymers and chemicals, nanotubes, and dendrimers in various applications besides water treatment. It has been anticipated that nanomaterials like silica, TiO_2, clays, metal powders, and polymers will be in great demand in the coming years. The role of CNTs, fullerenes, dendrimers, and membrane-derived nanomaterials for processes like RO, NF, and UF in wastewater treatment is formidable, with their advantages of flexibility, easier operation and maintenance, and scalability [43]. However, besides the many expectations, various challenges remain to be met, to bring them into the commercial market. The crucial aspects that must be thoroughly taken care of are: commercial availability of nanomaterials, extent of compatibility with the existing infrastructure, and associated health risks (such as leaching of nanomaterials into treated water). It has been found that the water generated directly from various sources (household and industrial) is rich in suspended particles that impede the treatment, particularly removal of soluble recalcitrant contaminants. It has been suggested to use nanotechnology-based water treatment methods followed by conventional treatment methods, which will lower the loads of suspended particles and help in efficient water treatment in a cost-effective way. Further, using nanotechnology either in the primary or secondary stages rather than in the final stages of water treatment would lower the risk of any nanomaterials leaching into the treated water and coming in direct contact with humans and the environment. Various other measures have been under consideration to make nanotechnology-based water treatment a complete success.

3.4 Enzyme technology and water remediation

Enzymes have high specificity (even a hundred or a thousand times higher than nanoparticles), high catalytic turnover, cost effectiveness, nontoxicity, and are nonerosive. They can degrade recalcitrant water pollutants in an ecofriendly manner, requiring no toxic chemicals for their action. They work under mild physical conditions, which makes the process energy saving. Some enzymes require additional factors for their action, particularly metal ions like Cu, Ni, Mg, Mn, Zn, etc. Enzymes can work in the presence of foams, sprays, lotions, and even detergents, which has made them even more in demand for water treatment. Generally, enzymes are extracted from microbial or plant sources. However, recombinant enzymes are also prepared when the source has insufficient enzyme concentrations. Most importantly, enzymes have been proven to degrade those recalcitrant water contaminants that cannot be degraded by oxidation (chemical, photochemical, electrochemical), ozonation, photolysis using H_2O_2 and O_3, corona process, TiO_2 photolysis, radiolysis, and or even adsorbed or filtered by known nanoparticles in some cases (Tables 3.4 and 3.5). There are various reasons for contaminant recalcitrance, the foremost being the presence of unusual substitutions with halides (Cl^- or Br^-), azo (−N=N) linkage, very large molecular size, and the presence of unusual bonds or highly condensed aromatic rings or presence of tertiary and quaternary carbon atoms. Enzymes can act specifically on the unusual bonds or substitutions and produce degradation. Further, enzymes do not increase BOD or chemical oxygen demand (COD) of water while causing degradation of water contaminants. The following paragraphs give a brief outline of the various aspects of enzymes being used during water treatment.

(i) Water Decontamination

Water can be decontaminated using a variety of enzymes from plants or microbial sources. A brief summary of various enzymes used in removal of pollutants is given in the following paragraphs.

- *Phenolic contaminants and related compounds*: These are generated during treatment of coal, petroleum refining, synthesis of resins and plastics, metal coating, wood preservation, textiles, dyes and other chemicals, mining and dressing, and pulp and paper, etc. [44]. Their usage has been strictly regulated in most developed countries. Various classes of enzymes being used in their degradation are:
  (a) Peroxidases: They are used for hydrolyzing aromatic contaminants with their action dependent on the presence of peroxides (H_2O_2). They can be obtained from a number of microorganisms as well as plants. Different peroxidases

Table 3.4 Water Pollutant Degrading Enzymes

Enzyme	Source	Pollutants
Peroxidase	Horseradish	Phenol, chlorophenol, anilines: degradation, phenol determination, Kraft effluent: decontamination
	Atromyces ramosus	Phenol, polyaromatics, herbicides: degradation, humic acid polymerization
	Plant materials	Water decontamination
Chloroperoxidase	*Caldariomyces funago*	Oxidation of phenolic compounds, chlorophenol detection
Lignin peroxidase	*Phanerochaete chrysopsorium*	Phenol, aromatic compounds: degradation, Kraft effluent: decontamination
	Chrysonilia sitophila	Kraft effluent: decontamination
Manganese peroxidase	*Phanerochaete chrysopsorium*	Phenol, lignin, pentachlorophenol, dyes: degradation
	Lentinula edodes	Chlorophenol, diuron: degradation
Tyrosinase	*Agaricus bisporus*	Catechol oxidation
Laccase	*Trametes hispida*	Dye decoloration
	Pyricularia oryzae	Azo-dye degradation
	Trametes versicolor	Textile effluent, chlorophenol, urea derivatives: degradation
	Plant materials	Chlorophenol degradation, xenobiotic binding to humus
	Pycnoporus cinnabarinus	Benzopyrenes degradation
Catechol dioxygenase	*Comamonas testosteroni*	Chlorophenol oxidation, diuron degradation
	Pseudomonas pseudoalcaligenes	Polychlorinated biphenyls, chlorothanes: degradation
Phenol oxidase	*Trametes versicolor*	Chlorinated compounds: degradation
	Thermoascus aurantiacus	Kraft effluent decontamination

(based on substrate or sources) that have been used for treatment of water pollutant treatment are listed here:

1. Horseradish peroxidase (HRP): It can hydrolyze toxic aromatic compounds like phenols, biphenols, anilines, hydroxyquinoline, arylamine, carcinogens (benzidines and naphthylamines), and related heteroaromatic compounds. Further, compounds that are not the substrates of HRP can also be treated, as the enzyme helps in forming insoluble precipitates while acting on its substrates followed by sedimentation and filtration of precipitates [45]. The enzyme has a broad range of physico-chemical parameters (pH, temperature, ionic strength, etc.), and thus it is one of the most effective enzymes in treatment of wastewater highly contaminated with pollutants.

Table 3.5 Summary of Enzymes Used in Biosensors for Detecting Various Water Contaminants

Pollutants	Enzymes
Heavy metals	
Mercury, cadmium, arsenic, copper, lead	Urease
Phenolic compounds	
Binary mixtures like phenol/chlorophenol, catechol/phenol, cresol/chlorocresol, and phenol/cresol	Laccase and tyrosinase
Phenol, *p*-cresol, *m*-cresol, catechol	Polyphenol oxidase
Pesticides	
Simazina	Peroxidase
Parathion	Parathion hydrolase
Paraoxon	Alkaline phosphatase
Carbaril	Acetyl cholinesterase
Herbicides	
2, 4-dicholorophenoxyacetic acid	Acetyl cholinesterase

2. Lignin peroxidase (LiP): It is also called ligninase and diarylopropane oxygenase. It is mainly produced by white-rot fungus (*Phanerochaete chrysosporium*). This enzyme has lower stability towards extreme physico-chemical conditions than HRP; however, it can hydrolyze a wide range of aromatic recalcitrant compounds like polycyclic aromatic and phenolic compounds in addition to its natural substrate (lignin) [46, 47].
3. Other peroxidases: Chloroperoxidase produced by *Caldariomyces fumago* is used to oxidize several phenolic compounds through oxygen transfer reactions. Manganese peroxidase produced by *P. chrysosporium* catalyzes oxidation of several monoaromatic phenols and aromatic dyes in the presence of Mn^{2+} [48, 49].
4. Use of plant material: Plant material containing peroxidases is being directly used in mineralizing phenolic compounds present in water as well as soil. Generally, plants' roots are best suited for mineralization of phenolic compounds. Peroxidases from tomato and water hyacinth plants have broad specificity for a variety of phenolic substrates while that from minced horseradish, potato, and white radish are specific and act only on 2, 4-dichlorophenol and related compounds [50].

(b) Polyphenol oxidases: These are helpful in decontamination of phenolic pollutants present in wastewater.

1. Tyrosinase: It is also called polyphenol oxidase, phenolase, and catecholase [51]. It removes phenols by its hydroxylation followed by production of *o*-quinones, which are insoluble in water hence easily collected through filtration [52]. The enzyme usually works in the concentration range from 0.01 to 1.0 $g\,dm^{-3}$ of phenol present in wastewater.

2. Laccase: It is produced by many fungi, but *Rhizoctonia praticola* produces a larger amount of the enzyme than any other known sources. This enzyme is not very specific towards its substrate, but it can act on a number of similarly looking phenols. During its action, it generates anionic free radicals, which help in precipitation of many recalcitrant pollutants present in wastewater [53].

- *Pulp and paper wastes*
  (a) Peroxidases: During wood pulping, 5%–8% (w/w) of residual modified lignin is produced by the Kraft process, which imparts a brown color to the pulp. This brown color is removed by bleaching agents (chlorine and chlorine oxides) at commercial scale. These chlorinated compounds are highly toxic and mutagenic and are released into the effluents, creating a major environmental hazard. Horseradish peroxidase and lignin peroxidase (from *P. chrysosporium*) are helpful in treating Kraft mill effluents, particularly in color removal and degradation of lignin by oxidizing aromatic units to cation radicals, which can be easily decomposed [54, 55].
  (b) Laccase: This enzyme is useful in treating effluent coming from bleaching industries and helps in removal of chlorophenols and chlorolignins [56]. It is an intracellular enzyme produced by the fungus *Coriolus versicolor.*
  (c) Cellulolytic enzymes: These include cellobiohydrolase, cellulase, and β-glucosidase, helpful in treatment of sludge generated from pulping and deinking operations. They are able to convert cellulosic sludge (60 kg ton^{-1} of pulp) generated from pulping into ethanol (higher commercial value). Further, they are not inhibited by high ink content (present in pulp) and their efficiency can be increased by adding surfactants [57].
- *Pesticides*: They are chemicals used against herbs, weeds, insects, and fungal pathogens for crop protection. The worst part is not their recalcitrance towards degradation but their usage in amounts several times higher than are actually required, due to ignorance. This leads to their enhanced concentration not only in the growing crop but also in the water table deep inside the soil. Further, pesticide-producing industries have other serious concerns due to their continuous disposal of chemical wastes into surface waters. Physical (such as incineration) and chemical treatment of pesticides are not effective and they produce hazardous by-products and are expensive [58]. Enzyme-based treatment of pesticides has given some hope; presently only an enzyme has been identified to be useful in pesticide detoxification, which is parathion hydrolase [59].
  (a) Parathion hydrolase (phosphotriesterase): It is produced by bacteria like *Pseudomonas* sp., *Flavobacterium* sp. and a recombinant *Streptomyces* [60]. It can hydrolyze organophosphate pesticides (major proportion of agricultural pesticides with the highest toxicity) like methyl and ethyl parathion, diazinon, fensulfothion, dursban, and coumaphos. Hydrolysis of the mentioned pesticides using the enzyme helps in their solubilization followed by their degradation using UV ozonation [61]. The enzyme parathion hydrolase has temperature and pH stability at 50°C and 5.5–10.0°C, respectively [62].

- *Cyanide wastes*: These are produced during industrial processes like production of synthetic fibers, rubber, chemical intermediates, pharmaceuticals, ore leaching, coal processing, and metal plating, as well as from food and feed production, due to the presence of cyanogenic glycosides in various crop materials. Further, they are produced during natural biological processes through enzymatic action by plants, microbes, and insects. Over 3 million tons of CN^- is being discharged by various processes yearly in drinking water throughout the world. CN^- is extremely toxic, as it is the metabolic inhibitor of crucial respiratory enzymes.
  (a) Cyanidase: It is produced by *Alcaligenes denitrificans* having K_m of $0.02\,mg\,dm^{-3}$ for CN^- with optimum pH range from 7.8 to 8.3 [63]. It can convert CN^- present in wastewater into ammonia and formate without being affected by common ions like Fe^{2+}, Zn^{2+} and Ni^{2+} or organic substrates (acetate, formamide, acetamide, and acetonitrile) present in wastewater. This enzyme has been found to be very useful in treating waste coming from the food industry based on debittering apricot seeds.
  (b) Cyanide hydratase (formamide hydrolyase): It is produced by fungi like *Gloeocercospora sorghi* and *Stemphylium loti* for the treatment of industrial waste rich in CN^- [64].
- *Food-processing wastes*: These are produced by the food-processing industry in the millions of tons every year. They are a major cause of the increase in water BOD and make the water unfit for drinking and other applications.
  (a) Proteases: They can be obtained by *Bacillus subtilis*, *Bacillus megaterium*, *Pseudomonas marinoglutinosa*, and *Acromonas hydrophila* [65]. They are used in solubilizing waste streams coming from the food industry processing of meat, fish, and livestock [66]. They are also helpful in processing waste coming from poultry slaughterhouses, particularly in processing of feathers. However, before protease treatment there is a pretreatment of waste feathers by using NaOH, followed by mechanical disintegration and then enzymatic treatment. The process produces lots of protein as end products, which can be used as a feed constituent.
  (b) Amylases: They are used in reducing BOD of wastewater and treatment of wastewater through saccharification and fermentation. They are found to be useful in the synthesis of bioplastics (photodegradable and biodegradable) using waste containing cheese whey, potato waste, etc. [67]. They are also used in the preparation of mulch films, compost bags, and in programmable fertilizer and pesticide delivery systems (protect from excessive pesticide runoff) using food wastes present in water [68, 69].
  (c) Other enzymes: Pectinesterase from *Clostridium thermosulfurogenes* and pectin lyase from *Clostridium beijerinckii* are used to degrade pectin (cell wall component of most fruits) present in wastewater into butanol (commercially valuable). The enzyme L-galactonolactone oxidase, from *Candida norvegensis*, is used to convert galactose (present in whey contaminated water) into L-ascorbic acid (commercially valuable) [44].

The billions of kilograms of whey (responsible for increasing BOD of wastewater) produced annually by dairy industries can be treated by the lactase enzyme [70]. Chitinase from *Serratia marcescens* QMB1466 is used for shrimp pretreatment, leading to its size reduction, deproteination and demineralization, followed by production of single cell protein (SCP). Further, it is used in treatment of shellfish waste rich in chitin content [71].

- *Solid waste and sludge treatment*: Solid waste and sludge are generally accompanied with lignocellulosic and cellulosic waste, which can be treated with an enzyme system containing endoglucanase, cellobiohydrolase, and cellobiase and turned into useful products like sugars, ethanol, biogas, and other energetic end products [72]. This enzyme system has been found to be useful in treating the organic fraction present in municipal solid wastes (MSW) into fermentable sugars like ethanol or butanol under anerobic conditions, which can then be collected for various other applications [73]. The enzyme system econase (endo-1, 4-β-D-glucanase, cellobiohydrolase and exo-1, 4-β-D-glucosidase) is also helpful in treating MSW [74]. Sludge dewatering is done using carbohydrase, lipase, and proteinase to reduce the volume of sewage sludge [75]. Peroxidases are also helpful in dewatering MSW sludge as well as in sedimentation of slimy material (very difficult to settle) by removal of water molecules and enhancement of mechanical binding of slime particles. Further, they also promote growth of microbes like algae and molds, which helps in additional aggregation and increased viscosity, leading to effective sedimentation of slimy material [76].
- *Removal of heavy metals*: Various industrial processes like mixing, smelting, electroplating, synthesis of pigment, nuclear power, defense, and fuel reprocessing produce a number of heavy metals like arsenic, copper, cadmium, uranium, lead, chromium, strontium, etc. discharged into water. Phosphatase containing cells of *Citrobacter* sp. are found to be useful in remediation of wastewater containing heavy metals. The remediation of heavy metal-containing wastewater by the enzyme phosphatase involves production of insoluble phosphate, followed by precipitation of almost all types of heavy metals, and collection. This enzyme has optimal pH stability in the range from 5 to 9, with temperature stability from 10°C to 40°C. The enzyme can severely be affected by the presence of high concentrations (>5 mM) of ions like Cl^- and CN^- [77–79].

(ii) Water Hardness

Hard water contains metallic ions like calcium, magnesium, aluminum, barium, iron, manganese, strontium, and zinc, which make it unfit for drinking and other daily activities. Further, it is also responsible for precipitate deposition in hot water pipes, heaters, boilers, kitchens, bathtubs, and other units. The two most common methods used for hard water treatment are lime soda and ion exchange softening. They are popular due to their cost effectiveness and faster action. Other methods include distillation, nanofiltration, electrodialysis, CNTs, capacitive deionization, and reverse osmosis. Lime soda and ion exchange softening methods also have drawbacks. In the case of lime soda, there is production of lots of sludge, which requires additional treatment by harsh chemicals followed by acidic treatment. In the case of lime soda, it generates lots of sodium content in the treated water, which is responsible for health issues like stroke, high blood pressure, hypertension, etc. [80].

Enzymes have provided solutions for removing water hardness without generating any side effects, and in a cost-effective manner. However, they have not yet landed in the commercial market, as large-scale testing in still in process. The laboratory set-up (diagrammatically shown in Fig. 3.3) contains an electrolytic cell containing an anode (immobilized glucose dehydrogenase) onto Nafion membrane and a cathode of Pt/C. A porous ion exchanger membrane is between the anodic and cathodic chambers. There is a central chamber with an inlet of hard water and an outlet for treated water without metallic ions. A solution of glucose prepared in phosphate buffer (pH 7.1) in addition to NAD^+ fills the anodic chamber while there is only a phosphate buffer with the same pH in the cathodic chamber. During hard water softening, the cathodic chamber exchanges cations (such as Ca^{2+}, Mg^{2+}) while the anodic chamber exchanges anions (such as Cl^-). This is carried out by oxidation of glucose into gluconic acid with the reduction of NAD^+ to $NADH+H^+$ by

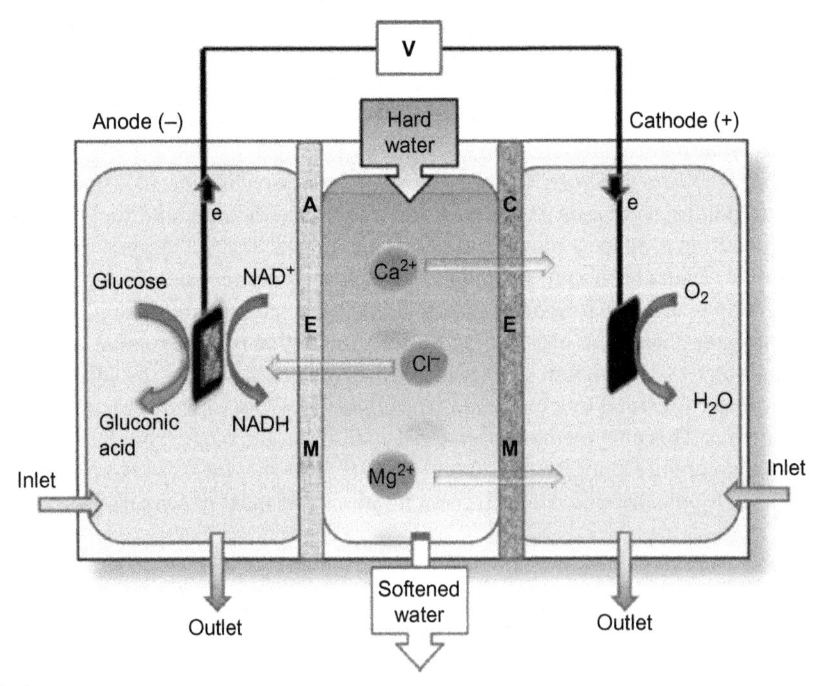

FIG. 3.3

Schematic of an enzymatic water softener (*AEM*, anion exchange membrane; *CEM*, cation exchange membrane).

Adapted from Nigam VK, Shukla P. Enzyme based biosensors for detection of environmental pollutants: a review. J Microbiol Biotechnol 2015;25:1773–1781.

the enzyme in the presence of oxygen being sparged along with the hard water. In the anodic chamber, production of H^+ ions leads to removal of Cl^- from hard water as HCl, while gluconate in the cathodic chamber combines with divalent (Ca^{2+} and Mg^{2+}), thus freeing the hard water from the ions responsible for its hardness. It has been found that Ca^{2+} can be removed more easily than Mg^{2+} as the latter has a larger hydrated radius. This set-up has an efficiency of 80%, which is highly dependent on the amount of Mg^{2+} in the hard water. The performance of this softener is dependent on the enzyme stability and efficiency of cofactor cycling. This type of enzyme softener has been recommended for residential use. Work is in process to enhance the enzyme stability by various methods, of which immobilization has been given utmost priority. The other enzyme substitute has also been looked into, to make the process NAD^+ independent, namely glucose oxidase [81].

(iii) Enzymes are Best Suited for Biofilm Removal

Biofilm is an extracellular polymeric substance (EPS) secreted by microbial species like *Leuconostoc mesenteroides*, *Pediococcus* sp., *Streptococcus salivarius*, *Sulfolobus acidocalcidarius*, *Bacillus substilis*, etc. [82]. It helps in cells' attachment to the surface, channeling of water and oxygen as well as nutrients, serving as a barrier against hostile environmental conditions, and providing resistance against antimicrobial agents. It can be either hydrophilic or hydrophobic, depending on environmental conditions. It contains polysaccharides (homo or heteropolysaccharides containing sugars like glucose, fructose, mannose, galactose, pyruvate, mannuronic acid, glucoronic acid, etc.), proteins (10–200kDa with ~40%–60% of hydrophobic amino acids), nucleic acids, lipids, and humic substances [83]. However, polysaccharides are the most important component as they provide adhesion to microbes on the surface. The composition of biofilm depends on microbial strain, stage of growth phase, limiting substrate (carbon, nitrogen, and phosphorus), oxygen limitation, ionic strength, temperature, and presence of shear force. A biofilm can be capsular (tight microbial binding) or slimy (loose microbial binding) depending on the constituents of the water. A biofilm is generally formed in water highly contaminated with pollutants rich in organic waste.

Biofilm has been a nuisance for water treatment and creates enormous problems for water treatment and drinking water distribution systems (increased concentration of contaminants), with reduced quality of potable water and increased rate of pipe corrosion. It leads to fouling of various instruments used during water treatment, followed by accelerated corrosion. Various methods have been tried for efficient biofilm removal including chemical (antibiotics, chemical biocides, detergents, surfactants, disinfectants like chlorine, chlorine dioxide) and physical methods (UV irradiation, scrabing, sonication, freezing and thawing). None of them are completely effective due to increased rates of genetic exchange, altered biodegradability, and increased secondary metabolite production. Further, these methods are time consuming, have poor penetrative ability, and are costly.

Enzymes have brought solutions due to their effective degradative capabilities, through direct action onto biofilm. The type of enzyme application requires complete information on the composition of the biofilm, which then determines the type of enzyme, its concentration, temperature, pH, ionic strength, etc. Enzymes act by destroying the physical integrity of biofilms by weakening of bonds present in their constituents like polysaccharides, proteins, lipids, nucleic acids, etc. It has been found that degradation of polysaccharides by polysaccharases has been very effective in degradation of biofilm, as it is the dominating factor for biofilm attachment and proliferation [84]. Enzymes like cellulase, α-amylase, and β-glucanase are effective biofilm removal agents, regardless of enzyme source. However, enzymes like protease, β-glucosidase, aminopeptidase, β-galactosidase, lipase, phosphatase, and lipase assist in biofilm degradation. Generally, bacterial and fungal sources are preferred due to ease of enzyme isolation and purification. The best recipe for enzyme preparation obtained *A. naeslundii, N. subflava, L. rhamnosus, P. gingivalis, S. oralis, S. mutans, V. dispar* and *S. sanguinis* [Patent CA2001539A1, Patent US4936994, Patent EP0388115A1] containing α-amylase, lipase, β-glucanase and protease enzymes, which have been found most effective with respect to any other combinations [85]. Protease present in the combination helps in protein degradation and makes it highly labile to be easily acted on by other polysaccharases, most importantly α-amylases. [86] Enzymatic action has been found to be more effective, more time saving, and less laborious in removal of biofilm than any other methods known currently. However, this method has not been yet commercialized due to lack of technical set-up and the presence of lower-cost biocides prevailing in the market. Enzyme technology through enzyme immobilization has been suggested, which would be a help in commercializing enzyme-based biofilm removal at large scale in water treatment plants.

(iv) Sensing and Monitoring

Enzymes are an excellent sensor in monitoring various water pollutants in treated water due to their catalytic specificity [87, 88]. The most commonly used enzyme-based biosensors are being used to monitor pollutants like carbamates, phosphates, organophosphates, alcohols, ammonia, cyanide, formaldehyde, organonitriles, phenol, zinc, and BOD as well as contamination by sewage. Enzymes are generally immobilized onto the transducer tip or entrapped inside a polymeric membrane made of poly-(2-glucosyloxyethyl methacrylate)-concanavalin, polyethylenimine, polyvinylferrocenium, etc. Sometimes, inorganic carriers like clay, porous silica, and alumina powder are used for enzyme immobilization, to provide better thermal and mechanical stability to the enzyme. They are preferred as they are nontoxic with respect to other reported carriers.

3.5 Combo-technology: water purification, disinfection, sensing, and monitoring

Enzymes are best suited for water remediation due to their ability to act on a vast variety of recalcitrant compounds. However, the soluble state of enzymes cannot be directly introduced into wastewater due to rapid inactivation and loss of catalytic efficiency [89]. A number of delivery systems can be used to introduce enzymes into wastewater, including:

- *Enzyme delivery by direct use of biological source*: Enzymes containing a source (microbial cells, plant tissue, or cells, etc.) are introduced directly into water effluent, rather than extracting soluble enzyme from the source. In the case of a plant source, either plant tissue or sometimes whole plants are used, depending on the enzyme secretion being used for degradation of pollutants. For example, degradation of dyes like malachite green, methyl orange, and brilliant blue R4 uses the enzyme peroxidase secreted from the roots of *Typhonium flagelliforme.* However, the major disadvantage of introducing plant parts or a complete plant is the increase in BOD and COD of the water effluent, which complicates the complete water remediation process. Enzymes from microbial cells are much more effective in water remediation, as enzymes can be induced to be synthesized by microbial cells depending on the type of pollutant. However, induction requires modification of the growth media with the addition of essential nutrients, which is generally not possible with in situ water remediation by enzymes. For example, a pure bacterial strain *Staphylococcus arlettae* can degrade azo dyes like CI Reactive Yellow 107, CI Reactive Red 198, CI Reactive Black 5, and CI Direct Blue 71 by 97%. However, degradation requires the presence of an active carbon source for effective pollutant degradation [90]. Further, enzymes containing microbial cells for pollutant degradation do not initiate as soon as microbial cells are introduced into the water. They need time for acclimatization to recalcitrant pollutants present in wastewater, which makes complete water remediation more time-consuming and less effective.
- *Enzyme delivery as cell-free enzyme extracts*: Cell extract containing the enzyme required for pollutant degradation is prepared through source lysis using chemical or physical methods. Here, the complexity of the growth medium as well as the time duration for acclimatization is not the issue being raised, thus making water remediation faster, simpler, and more cost effective. However, crude enzymes have lower catalytic efficiency than pure enzymes, in addition to the fact that pure enzymes are more susceptible to degradation by slight changes in the physico-chemical environment, which is not possible to control, particularly in wastewater heavily laden with recalcitrant pollutants [91].
- *Enzyme delivery in immobilized form*: This is the method best suited for introducing enzymes into wastewater without affecting enzyme conformation, an important factor in their catalytic activity. Enzymes are protected from extreme physico-chemical conditions like high temperature, very low or very high pH, high ionic strength, presence of inhibitors, etc., commonly found in polluted water. Further, this technique provides the additional benefit of increased reusability, making the complete process more cost effective with enhanced adaptability, easier handling, and higher specificity than any other known water treatment. Enzyme immobilization can be carried out by various means, such as adsorption, covalent binding, chemical coupling, etc. Enzyme matrices chosen for immobilization are well characterized with respect to their mechanical and hydrodynamic properties, filter characteristics, and biocidal properties, in addition to enzyme stability, reusability, and catalysis [44, 92–94].

Nanoparticles are best suited for acting as matrices for enzyme immobilization, due to their very large surface area to volume ratio, high reactivity, and sequestration properties (Fig. 3.4). Generally, enzymes used for water treatment are immobilized either by chemical coupling or covalent binding, as they provide increased reusability and more stability with minimal restrictions being imposed by pollutant diffusivity. However, there is a chance of enzyme inactivation when active sites are involved in enzyme attachment. In those cases, optimization is carried out with respect to the modifiers being used, and the conditions during enzyme immobilization are chosen with utmost care. Further, the enzyme immobilized by nanoparticles can be systematically moved from bench scale testing, followed by statistical analysis for lower volume systems, to commercialization for water treatment at pilot scale [95]. Thus complementing enzyme technology with nanotechnology would be very effective in water decontamination, due to removal of several drawbacks faced when used individually.

Several reports have been published on enzymes being immobilized onto nanoparticles for water pollutant degradation, sensing and removal of biofilm. Chymotrypsin (protease) has been immobilized inside a silicate shell-like cage (allowing pollutants to permeate) for biofilm degradation [96]; peroxidases, laccases, tyrosinase, dehalogenases, and organophosphorous hydrolases have been immobilized onto CNTs for degradation of various recalcitrant organic contaminants like phenols, polyaromatics, dyes, chlorinated compounds, and pesticides [97–101]. It has been found that nanoparticles in the form of nanosponges (made of nanopolymers) containing

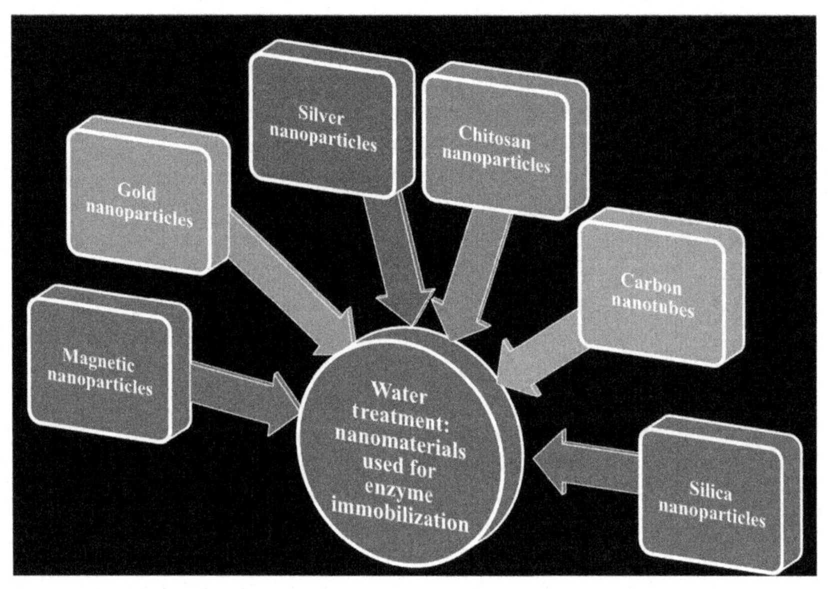

FIG. 3.4

An overview of various nanomaterials used for enzyme immobilization during effective water treatment.

microscopic particles with nanosized cavities are useful in enzyme immobilization by encapsulation, increasing the catalytic efficiency by severalfold [102]. Enzymes like peroxidase and laccase (used for water remediation containing waste coming from dyeing and textile-processing industries) have been effectively immobilized inside nanosponges [103]. Immobilized enzymes using nanoparticles have been very useful due to various abilities: hydrolysis of all types of recalcitrant and xenobiotic compounds, adaptability at high and low concentrations of contaminant, treatment can be done over a wide range of physico-chemical parameters (pH, temperature, salinity, etc.), high selectivity (can easily be used even in diluted water), negligible shock loading effects, negligible time for acclimatization, easier process control, less sludge volume due to the absence of biomass as well as inhibition by substances generally toxic to living organisms, cost effectiveness, and smaller retention times with respect to other treatment methods.

The most commonly used nanoparticles for enzyme immobilization are:

- *Magnetic nanoparticles*: These nanoparticles are much preferred over other nanoparticles for enzyme immobilization due to the presence of magnetism, which helps in easier separation of immobilized enzymes after water treatment. Most importantly, the superparamagnetic property of magnetic nanoparticles acts to retain magnetism in the presence of a magnetic field, while retaining no magnetism once the magnetic field is removed. Magnetic nanoparticles do not agglomerate and remain suspended in solution due to their supermagnetism, an important feature responsible for the excellent efficiency of immobilized enzymes. Further, the enzyme can be positioned accurately onto nanosized magnetic particles with the help of magnetic resonance imaging. In addition to providing enzyme stability and easier separation from the reaction mixture, magnetic nanoparticles are also used in sensing and monitoring various contaminants present in water. Generally, magnetic nanoparticles like FeO and Fe_2O_3 are used for enzyme immobilization due to size uniformity and excellent electrical, optical, magnetic, and chemical properties. Sometimes, magnetic nanoparticles are surface modified with polymers (dextran, polyvinyl alcohol; PVA, diethylaminoethyl; DEAE-starch), silica, etc. based on the enzyme type and application. Water-remediating enzymes like cellulase, catalase, β-glucosidase, lipase, peroxidase, agarase, κ-carrageenase, and laccase have been immobilized onto magnetic nanoparticles. Enzymes like alkaline phosphatase and glucose oxidase have been immobilized onto magnetic nanoparticles and are used for sensing and monitoring water contaminants [104–107].
- *Gold (Au) and silver (Ag) nanoparticles*: Besides providing a large surface area for enzyme immobilization, Ag and Au nanoparticles exhibit excellent electronic, optical, and thermal properties and act as conduction centers to facilitate transfer of the electrons. These nanoparticles are highly versatile with respect to their size and structure, which makes them best suited for enzyme immobilization in water treatment as well as in the synthesis of biosensors. It has been found that they provide favorable orientation of

enzyme immobilization (particularly redox enzymes) due to the presence of conducting channels between enzymes and the surface of nanoparticles. Thus they are helpful in imparting maximum stability and reusability to immobilized enzymes. However, they have a tendency of aggregating, due to having very high surface energy. Therefore, surface passivation of Ag and Au nanoparticles has been very important during enzyme immobilization. It can be done using citrate and thiol-functionalized organics, by self-assembling into monolayers, by encapsulation inside water pools of reverse microemulsions, etc. Generally, enzymes are immobilized onto Ag and Au nanoparticles using physical and covalent (direct and through cross-linker) adsorption. The covalent method is the more preferred, as it imparts better stability and versatility of enzyme immobilization. However, the covalent method often leads to enzyme inactivation as compared to the physical method. The covalent method involves functionalization of Ag and Au nanoparticles using cysteamine or cetyl trimethylammonium bromide (CTAB), followed by either direct enzyme addition or after addition of a spacer (glutaraldehyde). Spacer-based enzyme immobilization is more effective, as there is less chance of enzyme inactivation and this technique imparts better catalytic efficiency. There should be intensive optimization, particularly the ratio of functionalization reagent and spacer, to minimize excessive aggregation and enzyme inactivation. CTAB-based enzyme immobilization has been generally used for water-treating enzymes; however, it requires a larger amount of enzyme for immobilization. Water-treating enzymes that are immobilized onto Ag and Au nanoparticles are: lipase, peroxidase, keratinase, α-amylase, esterase, urease, endoglucanase, carboxypeptidase, and laccase. Further, sensing and monitoring enzymes immobilized onto Ag and Au nanoparticles are: choline oxidase, glucose oxidase, and tyrosinase [108–111].

- *Chitosan nanoparticles*: Chitosan is a hydrophilic, positively charged biopolymer with uniform dispersion, molecular mobility, relaxation behavior, and adaptable thermal and mechanical properties. These nanoparticles have size versatility ranging from a nanometer to a millimeter, which is helpful in immobilizing variety of enzymes as well as also helpful in types of applications. However, the smaller the size of the chitosan nanoparticle, the greater is its immobilization efficiency. Further, they do not have any magnetic properties and have higher fragility with respect to any nanoparticles being used for enzyme immobilization. Enzyme immobilization onto chitosan nanoparticles is much easier without requirement of any additional chemical reagent due to presence of number of active functional groups. There is proper distribution of enzyme molecules onto chitosan nanoparticles due to regular arrangement of active functional groups onto the surface of chitosan nanoparticles. Enzymes immobilized onto chitosan nanoparticles are highly stable towards adverse physicochemical conditions (pH, temperature, ionic strength), resistant to proteases and various denaturing compounds. Chitosan nanoparticles provide an ideal microenvironment for enzyme catalysis in water treatment. Various water-treating enzymes immobilized onto chitosan nanoparticles are: lipase, laccase, peroxidase, invertase, cellulase, esterase, phytase, neutral proteinase, keratinase, glucoamylase, and α-amylase. A sensing and monitoring enzyme immobilized onto chitosan nanoparticles is alkaline phosphatase [112–116].

- *CNTs*: These form one of the excellent matrices for enzyme immobilization due to their cylindrical shape, with the length in micrometers while the diameter is 100 nm. The surface of the nanotubes can be single layered (SWCNT) or multilayered (MWCNT), which provides better versatility to enzyme immobilization. However, SWCNTs are preferred, as they provide a larger surface area for enzyme interaction, while MWCNTs are cost effective and have easier dispersibility. CNTs provide high mechanical strength, are chemically inert, have excellent thermal conductivity, and are environmentally stable. Enzyme immobilization requires chemical functionalization by oxidizing sidewalls using HNO_3, H_2SO_4, $KMnO_4$, O_3 which helps in introducing functional groups like carboxylic acid, hydroxyl, phenol, carbonyl and quinone groups onto the nanotube's surface. These functional groups are used for covalent enzyme binding through esterification, thiolation, alkylation, and arylation. In some cases, enzyme molecules are trapped inside the hollow channel of the nanotube by simple adsorption. CNTs can also be modified with magnetic nanoparticles or complex (chitosan and magnetic nanoparticles) to attain their additional qualities, and it is also helpful in preventing nanoparticle aggregation after enzyme immobilization inside reaction mixtures. Water-treating enzymes that are immobilized onto CNTs are: lipase, laccase, peroxidase, and cellulase. Further, a sensing and monitoring enzyme immobilized onto CNTs is glucose oxidase. Sometimes a specific protocol has to be designed for a particular enzyme immobilization, for example, enzyme peroxidase can be actively immobilized onto MWCNTs by covalent linkage onto a functionalized nanotube surface with cross-linker (glutaraldehyde), followed by blocking of nonbound functional groups using *N*-hydroxysuccinimide (NHS) or *N*-ethyl-*N*′-3-dimethylaminopropylcarbodiimide hydrochloride (EDC). Immobilized peroxidise onto MWCNTs are found to be highly stable and catalytically active for the degradation of recalcitrant compounds present in waste water. Treated water can be monitored with immobilized enzymes (glucose oxidase and peroxidase) onto complex chitosan-SWCNTs [117–121].
- *Silica nanoparticles*: Enzymes are immobilized onto silica nanoparticles via covalent linkage using silanol groups present on their surface. Silica nanoparticles impart excellent enzyme stability and are less toxic due to fabrication. During enzyme immobilization, silanol groups present on the surface of silica nanoparticles are functionalized with 3-aminopropyltrimethoxysilane to generate -NH_2 followed by addition of glutaraldehyde (act as a cross-linker), which provides aldehyde groups to enzymes as well as aminolyted silica nanoparticles containing -NH_2 and forms imine bonds in both cases. The size of silica nanoparticles can easily be controlled based on synthesis protocols; they are biocompatible and provide a

very large surface area with an abundance of sites for surface functionalization. These features are responsible for excellent immobilization efficiency and better catalysis by immobilized enzymes. In addition, immobilized enzymes are less susceptible to damage by adverse physical and chemical conditions like pH, temperature, organic solvents, ionic strength, etc. Further, a minimal amount of aggregation has been observed in the case of immobilized enzymes kept in the reaction mixture for longer duration. Water-treating enzymes immobilized onto silica nanoparticles are lipase, peroxidase, α-amylase, and keratinase; while sensing and monitoring enzymes immobilized onto silica nanoparticles are catalase and glucose oxidase [122–125].

3.6 Combo-technology: bridging the gap between market and bench work

Water treatment using nanoparticles with attached enzymes (combo-technology) has been found to be more efficient and less toxic. Lab experimentation has found this technology to be very useful in effective removal of recalcitrant pollutants and combining the two is complementary. However, the combo-technology has not yet arrived on the market due to various challenges. The foremost of these are scale-up optimization, unstable physical states due to aggregation, and determination of an accurate number of cycles of reusability. Nanoparticles have gained more attention in enzyme immobilization, as they present a very high surface-to-volume ratio, leading to very high immobilization efficiency. The major problem that has been frequently observed is that immobilized enzyme on nanoparticles has a higher tendency of aggregating during storage. Microparticles with larger size, though not having higher immobilizing efficiency, at least have a tendency to aggregate due to less surface energy [126]. The optimal conditions under which aggregation of immobilized enzyme onto nanoparticles can be minimized have not yet been determined. Aggregation leads to heterogeneity and lowers the efficiency of water treatment. It has been found that removal of immobilized enzyme from the reaction medium after catalysis can help in minimizing aggregation. This creates a nuisance, however, as separation of immobilized enzyme (onto nanoparticles) is not an easy job due to the very small size. The concept is then to look for those nanoparticles that have less surface energy and can be separated due to added properties like magnetism. Enzymes immobilized onto magnetic nanoparticles can easily be separated from the reaction medium with the application of a magnetic field. Further, magnetic nanoparticles have less tendency to aggregate when the enzyme covers their surface by immobilization.

However, magnetic nanoparticles cannot be used for immobilization of all types of enzymes. Chitosan nanoparticles have been found to be very useful in this respect; however, they themselves have no characteristics for taking part in any water treatment. Therefore, hybrid magnetic chitosan nanoparticles have been made containing a magnetic coating (such as Fe_3O_4-chitosan nanoparticles) by using methods like microemulsion polymerization, or in situ polymerization. They have a very small tendency of aggregation and present very high immobilization efficiency. Further, they provide excellent enzyme stability with higher number of reusability cycles [127]. Magnetic chitosan nanoparticles are resistant to corrosion and offer higher flexibility with respect to functional groups. It has been found that they are a better sensor for monitoring the level of recalcitrant pollutants in treated water [128]. Similar types of nanoparticles are being explored that can fulfill all the requirements mentioned previously.

Water treatment using combined technologies of nanotechnology and enzyme technology has revolutionized both fields, with more advantages than their individual counterparts. They have exceeded the performance of conventional methods for removal of complex and recalcitrant pollutants in a less expensive manner, but there is still a long way to go before these products are accepted in the commercial market. Toxicological studies must be done despite claims of the nontoxicity of enzyme-immobilized nanoparticles with experimental evidence. The phenomenon of nanoparticle aggregation has to be properly taken care of, as it is one of the major reasons for biological system toxicity. Nanoparticle aggregation has been a major reason for disruption of the aquatic food chain, thus playing an important role in the collapsing aquatic ecosystem. Therefore, eco-friendly water treatment would be more effective, sustainable, and acceptable by every section of society.

References

[1] Muenchhoff M, Goulder PJR. J Infect Dis 2014;209:S120–6.

[2] Polimeni JM, Almalki A, Iorgulescu RI, Albu L-L, Parker WM, Chandrasekara R. Int J Environ Res Public Health 2016;13:1181.

[3] Gehrke I, Geiser A, Somborn-Schulz A. Nanotechnol Sci Appl 2015;8:1–17.

[4] Semisch A, Ohle J, Witt B, Hartwig A. Part Fibre Toxicol 2014;11:1–16. https://doi.org/10.1186/1743-8977-11-10.

[5] Ramasamy M, Lee J. Biomed Res Int 2016;2016:1851242.

[6] Keller VDJ, Williams RJ, Lofthouse C, Johnson AC. Environ Toxicol Chem 2014;33:447–52.

[7] Bhanja SN, Mukherjee A, Rodell M, Wada Y, Chattopadhyay S, Velicogna I, Pangaluru K, Famiglietti JS. Sci Rep 2017;7:7453.

[8] Zessner M, Lampert C, Kroiss H, Lindtner S. Water Sci Technol 2010;62:223–30.

[9] Lam S, Nguyen-Viet H, Tuyet-Hanh TT, Nguyen-Mai H, Harper S. Int J Environ Res Public Health 2015;12:12863–85.

[10] Dickin SK, Schuster-Wallace CJ, Qadir M, Pizzacalla K. Environ Health Perspect 2016;124:900–9.

[11] Rodriguez C, Buynder PV, Lugg R, Blair P, Devine B, Cook A, Weinstein P. Int J Environ Res Public Health 2009;6:1174–209.

[12] Lu S, Wang J, Pei L. Int J Environ Res Public Health 2016;13:298.

[13] Gumisiriza R, Hawumba JF, Okure M, Hensel O. Biotechnol Biofuels 2017;10:11.

[14] Jimenez B, Barrios JA, Mendez JM, Diaz J. Water Sci Technol 2004;49:251–8.

[15] Kumar GS, Kar SS, Jain A. Indian J Occup Environ Med 2011;15:93–6.

[16] Chaturvedi S, Dave PN, Shah NK. J Saudi Chem Soc 2012;16:307–25.

[17] Westerhoff P, Alvarez P, Li Q, Gardea-Torresdey J, Zimmerman J. Environ Sci Nano 2016;3:1241–53.

[18] Sperling RA, Parak WJ. Phil Trans R Soc A 2010;368:1333–83.
[19] Mylvaganam K, Zhang LC. Nanotechnology 2007;18:475701–4.
[20] Han J, Gao C. Nano-Micro Lett 2010;2:213–26.
[21] Wang X, Guo Y, Yang L, Han M, Zhao J, Cheng X. J Environ Anal Toxicol 2012;2:1–7.
[22] Chowdhury S, Balasubramanian R, Das P. Green chemistry for dyes removal from wastewater. Hoboken, NJ: John Wiley & Sons, Inc.; 201535–82.
[23] Machado FM, Fagan SB, da Silva IZ, de Andrade MJ. Carbon nanomaterials as adsorbents for environmental and biological applications. Switzerland: Springer International Publishing; 201511–32.
[24] Li J, Chen C, Zhang S, Wang X. Environ Sci Nano 2014;1:488–95.
[25] Gao W, Majumder M, Alemany LB, Narayanan TN, Ibarra MA, Pradhan BK, Ajayan PM. ACS Appl Mater Interfaces 2011;3:1821–6.
[26] Ray PZ, Shipley HJ. RSC Adv 2015;5:29885–907.
[27] Kharissova OV, Dias HVR, Kharisov BI. RSC Adv 2015;5:6695–719.
[28] Sivashankar R, Sathya AB, Vasantharaj K, Sivasubramanian V. Environ Nanotechnol Monit Manag 2014;1:36–49.
[29] Abbasi E, Aval SF, Akbarzadeh A, Milani M, Nasrabadi HT, Joo SW, Hanifehpour Y, Nejati-Koshki K, Pashaei-Asl R. Nanoscale Res Lett 2014;9:1–10.
[30] Nataraj SK, Hosamani KM, Aminabhavi TM. Water Res 2006;40:2349–56.
[31] Liu M, Lü Z, Chen Z, Yu S, Gao C. Desalination 2011;281:372–8.
[32] Wang L-S, Gupta A, Rotello VMACS. Infect Dis Ther 2016;2:3–4.
[33] Wang X, Hsiao BS. Curr Opin Chem Eng 2016;12:62–81.
[34] Yin J, Deng B. J Membr Sci 2015;479:256–75.
[35] Song X, Wang L, Mao L, Wang Z. ACS Sustain Chem Eng 2016;4:2990–7.
[36] Jhaveri JH, Murthy ZVP. Desalin *Water Treat* 2016;57:26803–19.
[37] Wagh P, Parungao G, Viola RE, Escobar IC. Sep Purif Technol 2015;156:754–65.
[38] Lewis SR, Datta S, Gui M, Coker EL, Huggins FE, Daunert S, Bachas L, Bhattacharyya D. Proc Natl Acad Sci U S A 2011;108:8577–82.
[39] Henderson MA, Lyubinetsky I. Chem Rev 2013;113:4428–55.
[40] Lazar MA, Varghese S, Nair SS. Recent Updates Catal 2012;2:572–601.
[41] Miranda ML, Kim D, Hull AP, Paul CJ, Galeano MAO. Environ Health Perspect 2007;115:221–5.
[42] Kaittanis C, Santra S, Perez JM. Adv Drug Deliv Rev 2010;62:408–23.
[43] Dimitrijevic J, Krapf L, Wolter C, Schmidtke C, Merkl J-P, Jochum T, Kornowski A, Schüth A, Gebert A, Hüttmann G, Vossmeyer T, Weller H. Nanoscale 2014;6:10413–22.
[44] Karam J, Nicell JA. J Chem Technol Biotechnol 1997;69:141–53.
[45] Liu W, Wang WC, Li HS, Zhou X. Water Sci Technol 2011;63:1621–8.
[46] Haritash AK, Kaushik CP. J Hazard Mater 2009;169:1–15.
[47] Stolz A. Appl Microbiol Biotechnol 2001;56:69–80.
[48] Rotta CEL, D'Elia E, Bon EPS. Electron J Biotechnol 2007;10:1–10.
[49] Salvachúa D, Prieto A, Martínez AT, Martínez MJ. Appl Environ Microbiol 2013;79:4316–24.
[50] González PS, Agostini E, Milrad SR. Chemosphere 2008;70:982–9.
[51] https://ghr.nlm.nih.gov/gene/TYR.
[52] Seetharam GB, Saville BA. Water Res 2003;37:436–40.
[53] Ma H-L, Kermasha S, Gao J-M, RM Borges X-ZY. J Mol Catal B Enzym 2009;57:89–95.
[54] Kersten PJ, Kalyanaraman B, Hammel KE, Reinhammar B, Kirk TK. Biochem J 1990;268:475–80.

[55] Royer G, Yerushalmi L, Rouleau D, Desrochers M. J Ind Microbiol 1991;7:269–77.
[56] Upadhyay P, Shrivastava R, Agrawal PK. 3 Biotech 2016;15:1–12.
[57] Sewalt VJH, Glasser WG, Beauchemin KA. J Agric Food Chem 1997;45:1823–8.
[58] Gupta VK, Ali I, Saleh TA, Nayak A, Agarwal S. RSC Adv 2012;2:6380–8.
[59] Nannipieri P, Bollag JM. J Environ Qual 1990;20:510–7.
[60] Rowland SS, Speedie MK, Pogell BM. Appl Environ Microbiol 1991;440–4.
[61] Grice KJ, Payne GF, Karns JS. J Agric Food Chem 1996;44:351–7.
[62] Wang XX, Chi Z, Ru SG, Chi ZM. Biodegradation 2012;23:763–74.
[63] Ingvorsen K, Højer-Pedersen B, Godtfredsen SE. Appl Environ Microbiol 1991; 57:1783–9.
[64] Kumar V, Kumar V, Bhalla TC. 3 Biotech 2015;5:641–6.
[65] Zhou M-Y, Chen X-L, Zhao H-L, Dang H-Y, Luan X-W, Zhang X-Y, He H-L, Zhou B-C, Zhang Y-Z. Microb Ecol 2009;58:582–90.
[66] Kim WK, Lorenz ES, Patterson PH. Poult Sci 2002;81:95–8.
[67] Song JH, Murphy RJ, Narayan R, Davies GB. Philos Trans R Soc Lond Ser B Biol Sci 2009;364:2127–39.
[68] Haight GP, Jursich GM, Kelso MT, Merrill PJ. Inorg Chem 1985;24:2740–6.
[69] Roy A, Singh SK, Bajpai J, Bajpai AK. Cent Eur J Chem 2014;12:453–69.
[70] Dwevedi A, Kayastha AM. J Plant Biochem Biotechnol 2010;19:9–20.
[71] Wang S, Hwang J. Enzym Microb Technol 2001;28:376–82.
[72] Sun Y, Cheng J. Bioresour Technol 2002;83:1–11.
[73] Lagerkvist A, Water CH. Sci Technol 1993;27:47–56.
[74] Bonilla S, Tran H, Allen DG. Water Res 2015;68:692–700.
[75] Green EM. Curr Opin Biotechnol 2011;22:337–43.
[76] Auriol M, Filali-Meknassi Y, Adams CD, Tyagi RD, Noguerol T-N, Piña B. Chemosphere 2008;70:445–52.
[77] Bautista FM, Bravo MC, Campelo JM, Garcia A, Luna D, Marinas JM, Romero AA. J Mol Catal B Enzym 1999;6:473–81.
[78] Greenwood JM, Gilkes NR, Kilburn DG, Miller RC, Warren RA. FEBS Lett 1989;244:127–31.
[79] Beddows CG, Gil HG, Guthrie JT. Biotechnol Bioeng 1985;27:579–84.
[80] Ha SK. Electrolyte Blood Press 2014;12:7–18.
[81] Nigam VK, Shukla P. Enzyme based biosensors for detection of environmental pollutants: a review. J Microbiol Biotechnol 2015;25:1773–81.
[82] Arugula MA, Brastad KS, Minteer SD, He Z. Enzym Microb Technol 2012;51:396–401.
[83] Donlan RM. Emerg Infect Dis 2002;8:881–90.
[84] Fletcher M, Loeb GI. Appl Environ Microbiol 1979;37:67–72.
[85] Stiefel P, Mauerhofer S, Schneider J, Maniura-Weber K, Rosenberg U, Ren Q. Antimicrob Agents Chemother 2016;60:3647–52.
[86] Kaplan JB. Methods Mol Biol 2014;1147:203–13.
[87] Justino CIL, Duarte AC, Rocha-Santos TAP. Sensors (Basel) 2017;17:2918.
[88] Kimmel DW, LeBlanc G, Meschievitz ME, Cliffel DE. Anal Chem 2012;84:685–707.
[89] Vu B, Chen M, Crawford RJ, Ivanova EP. Molecules 2009;14:2535–54.
[90] Mugdha A, Usha M. Sci Rev Chem Commun 2012;2:31–40.
[91] Kagalkar AN, Jagtap UB, Jadhav JP, Govindwar SP, Bapat SA. Planta 2010;232:271–85. https://doi.org/10.1007/s00425-010-1157-2.
[92] Vijaykumar MH, Veeranagouda Y, Neelkanteshwar K, Karegoudar TB. World J Microbiol Biotechnol 2006;22:157–62.

[93] Dellamatrice PM, Monteiro RTR. Quím Nova 2006;29:419–21.
[94] Singh N, Singh J. Prep Biochem Biotechnol 2002;32:127–33.
[95] Bolibok P, Wiśniewski M, Roszek K, Terzyk AP. Naturwissenschaften 2017;104:36.
[96] Misson M, Zhang H, Jin B. J R Soc Interface 2015;102:20140891.
[97] Majeau JA, Brar SK, Tyagi RD. Bioresour Technol 2010;101:2331–50.
[98] Bansal N, Kanwar SS. Sci World J 2013;2013:714639.
[99] Dvorak P, Bidmanova S, Damborsky J, Prokop Z. Environ Sci Technol 2014;48:6859–66.
[100] Kim I, Kim GH, Kim CS, Cha HJ, Lim G. Sensors (Basel) 2015;15:12513–25.
[101] Zhou Y, Zhang F, Tang L, Zhang J, Zeng G, Luo L, Liu Y, Wang P, Peng B, Liu X. Sci Rep 2017;7:43831.
[102] Osma JF, Toca-Herrera JL, Rodríguez-Couto S. J Enzym Res 2010;2010:918761.
[103] Ba S, Kumar VV. Crit Rev Biotechnol 2016;36:819–32.
[104] Gupta MN, Kaloti M, Kapoor M, Solanki K. Artif Cells Blood Substit Immobil Biotechnol 2011;39:98–109.
[105] Bergemann C, Muller-Schulte D, Oster J, Brassard LA, Lubbe AS. J Magn Magn Mater 1999;194:45–52.
[106] Wang S, Su P, Huang J, Wu J, Yang Y. J Mater Chem B 2013;1:1749–54.
[107] Huang J, Zhao R, Wang H, Zhao W, Ding L. Biotechnol Lett 2010;32:817–21.
[108] Petkova GA, Záruba K, Žvátora P, Král V. Nanoscale Res Lett 2012;7:1–10.
[109] Upadhyay LS, Verma N. Bioprocess Biosyst Eng 2014;37:2139–48.
[110] Lan DD, Li BB, Zhang ZZ. Biosens Bioelectron 2008;24:940–4.
[111] Jain P, Pradeep T. Biotechnol Bioeng 2005;90:59–63.
[112] Kayastha AM, Srivastava PK. Appl Biochem Biotechnol 2001;96:41–53.
[113] Zhao L-M, Shi L-E, Zhang Z-L, Chen J-M, Shi D-D, Yang J, Tang Z-X. Braz J Chem Eng 2011;28:353–62.
[114] Zang L, Qiu J, Wu X, Zhang W, Sakai E, Wei Y. Ind Eng Chem Res 2014;53:3448–54.
[115] Malmiri HJ, Jahanian MAG, Berenjian A. Am J Biochem Biotechnol 2012;8:203–19.
[116] Sowjanya NT, Dhivya R, Meenakshi K, Vedhanayakisri KA. Res J Eng Technol 2013;4:288–94.
[117] Saifuddin N, Raziah AZ, Junizah AR. J Chem 2013;2013:1–18.
[118] Slaughter G, Kulkarni T. J Biochip Tissue Chip 2015;5:1–10. https://doi.org/10.4172/2153-0777.1000110.
[119] Kim BJ, Kang BK, Bahk YY, Yoo KH, Lim KJ. Curr Appl Phys 2009;9:263–5.
[120] Oliveira SF, Bisker G, Bakh NA, Gibbs SL, Landry MP, Strano MS. Carbon 2015;95:767–79.
[121] Tîlmaciu C-M, Morris MC. Front Chem 2015;3:1–21.
[122] Rao KS, El-Hami K, Kodaki T, Matsushige K, Makino K. *J Colloid* Interface Sci 2005;289:125–31.
[123] Banjanac K, Mihailović M, Prlainović N, Stojanović M, Carević M, Marinković A, Bezbradica D. J Chem Technol Biotechnol 2016;91:439–48.
[124] Banjanac K, Mihailović M, Prlainović N, Ćorović M, Carević M, Marinković A, Bezbradica D. J Chem Technol Biotechnol 2016;91:2654–63.
[125] Cruz JC, Würges K, Kramer M, Pfromm PH, Rezac ME, Czermak P. Methods Mol Biol 2011;743:147–60.
[126] Müller KH, Motskin M, Philpott AJ, Routh AF, Shanahan CM, Duer MJ, Skepper JN. Biomaterials 2014;35:1074–88.
[127] Long J, Wu Z, Li X, Xu E, Xu X, Jin Z, Jiao A. J Agric Food Chem 2015;63:3534–42.
[128] Wang XY, Jiang XP, Li Y, Zeng S, Zhang YW. Int J Biol Macromol 2015;75:44–50.

CHAPTER

Production of clean air using combo-technology 4

Alka Dwevedi[*,†], **Jaigopal Sharma**[*]

Department of Biotechnology, Delhi Technological University, New Delhi, India[*]

Swami Shraddhanand College, University of Delhi, New Delhi, India[†]

4.1 Introduction

Polluted air has become a major health hazard worldwide, with developing countries being the topmost target. A number of technologies are available for monitoring air pollution and measuring its health effects. Air pollution has been a major cause of cardiopulmonary and respiratory diseases, and it has also been linked to lung cancer from exposure to elevated levels of atmospheric particulate matter (PM) [1]. Developing countries are lacking updated technologies for monitoring air pollution, so most of the toxic air pollutants go unchecked and lead to potential health risks. WHO (World Health Organization) in collaboration with the University of Bath (United Kingdom) has collected data around the world on air quality using satellite measurements, air transport models, and ground station monitors for >3000 locations, including both rural and urban. It was found that there is an increase of over 6% of air pollutants per year and this will continue to increase if it remains unchecked. The major factors in this increase are geographic and atmospheric conditions, scale and composition of economic activity, population, strength of local pollution regulations, and the energy mix [2]. It has been confirmed by WHO that 92% of the world's population lives in regions with excessive air pollutants and PM. Further, it has also been pointed out that air pollutants are present not only outside but they are found in great numbers inside homes and other buildings. There are over three million deaths (about 11.6% of global deaths) every year due to air pollution associated with both indoor and outdoor pollution [3] (Fig. 4.1). Nearly 90% of air pollution–related deaths have been found to occur in low- and middle-income countries. Further, nearly two out of three of these countries are in Southeast Asia and Western Pacific regions. Rich countries have taken a number of measures to control air pollution; however, they are not completely unaffected. Europe has suffered from high levels of ammonia and methane gases generated from diesel-powered cars and farming policies; the United States has been affected by ozone, etc. Women, children, and older adults are the most vulnerable targets of air pollution [4]. The most common causes of air pollution are inefficient transportation modes, household fuel and waste

Solutions to Environmental Problems Involving Nanotechnology and Enzyme Technology
https://doi.org/10.1016/B978-0-12-813123-7.00004-9

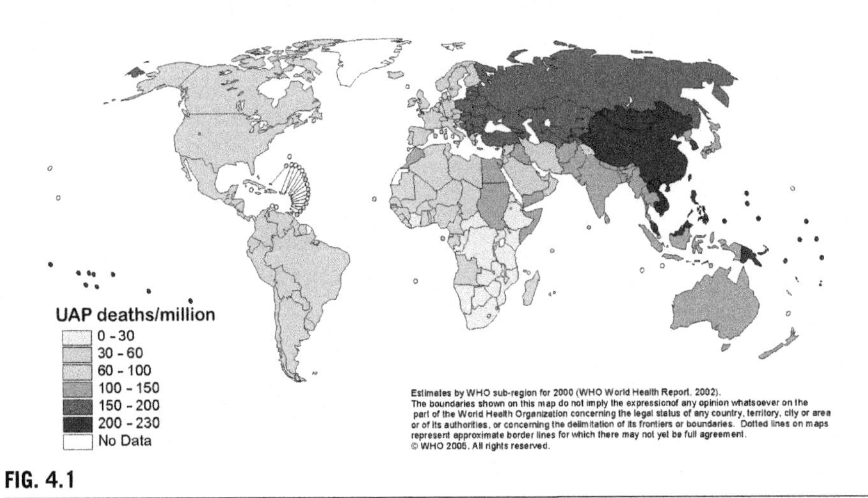

FIG. 4.1

Deaths worldwide due to urban air pollution (UAP).

WHO World Health Report.

burning, coal-fired power plants, and industrial activities. Further, air quality can also deteriorate due to dust storms (in desert and near-desert regions).

It has been found that in today's world about 90% of our time is spent indoors. Therefore we are more prone to effects from indoor air pollutants. Indoor air is far more polluted than outdoor air, due to the more limited oxygen supply (every man, woman, and child exchanges between 10,000 and 70,000 L of air every 24 h to sustain life). According to the EPA (Environmental Protection Agency), there are two to five times more pollutants present in indoor air, which sometimes increases to 100 times for various reasons [5]. Health problems like asthma, allergies, other respiratory problems, headaches, eye and skin irritations, sore throat, colds and flu, memory loss, dizziness, fatigue, and depression are frequently found during exposure to indoor pollutants [6].

WHO has given guideline limits for an annual mean of $PM_{2.5}$ at $10\,\mu g\,m^{-3}$. $PM_{2.5}$ includes air pollutants such as sulfate, nitrates, mineral dust, ammonia, and black carbon, which pose the greatest health risks, particularly to lungs and the cardiovascular system [7]. $PM_{2.5}$ exposure above $10\,\mu g\,m^{-3}$ leads to tissue and systemic inflammation, increases oxidative damage to DNA and cell membrane lipids, increases the risk for thrombosis, lowers birth weight, and impairs metabolic, cognitive, and immune function, causes developmental delays in children, leads to premature death, causes reproductive health problems, etc. [8]. Therefore, there is an urgent need to control air pollution in addition to efforts at the personal and community level, such as sustainable transport in cities, solid waste management, access to clean household fuels and cookstoves, as well as renewable energies and industrial emissions reductions, etc.

The most commonly known methods for controlling air pollution, including bag house, electrostatic precipitation, flue gas desulfurization (FGD), fabric filters, mercury control, and cyclone collectors, are based on air filtration and reducing PM via precipitation. Scrubbers and pressure swing adsorption (PSA) are typically used for removal of CO_2 via adsorption. These methods require intensive labor, costly installations, and a very large area for their operation [9]. The air purification technologies available on the market are summarized in Fig. 4.2.

Nanotechnology has been helpful in controlling air pollution in several ways. For example, nanocatalysts (viz. nanofiber of manganese oxide) are being used to convert toxic gases into harmless gases much faster, due to their large surface area. Nanostructured membranes with predefined pore size are used to trap harmful gases. The NANOGLOWA project has developed a range of nanostructured membranes to replace scrubbers and significantly reduce CO_2 in emissions. GE has developed nanostructured membranes helpful in effective capturing of CO_2 used as fuel. Carbon nanotubes (CNTs) have gained in significance due to their several hundred times higher capacity of trapping harmful gases and being the best choice in industrial scale plants for air purification [10]. Further, CNTs are being used to detect levels of gases like NO_2, NH_3, or O_3 based on changes in electrical resistance induced by charge transfer with gas molecules due to physical adsorption. Further, combining nanotechnology with enzyme technology has enhanced the air-purifying capabilities and sensing of toxic air pollutants by several thousand folds. CNTs with immobilized enzymes have been found to establish fast electron transfer from enzyme active sites to the electrode, leading to enhanced capability of detecting toxic gases in the air. Further, immobilized enzyme on CNTs has been manufactured by NanoTwin Technologies Inc. and commercialized under the trade name of Nanobreeze Room Air Purifier for air purification. The enzyme-like carbonic anhydrase has been immobilized onto various nanoparticles and used as the scrubber, converting CO_2 into carbonate in water solution, which can be used for various other applications. The

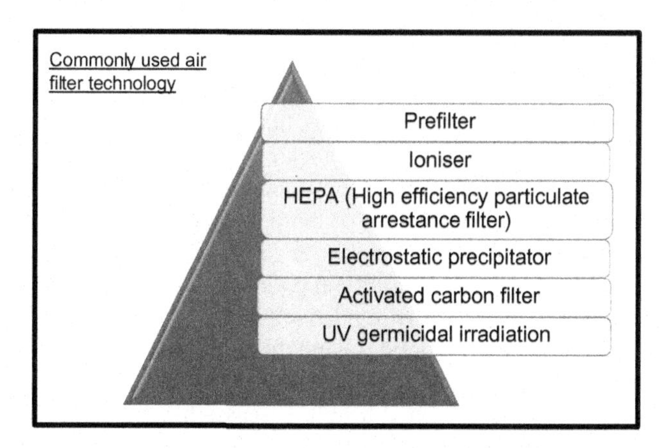

FIG. 4.2

Most commonly used air filter technology in the world.

photoactive nanomaterial TiO_2 has been used to trap nitrogen oxides in the presence of sunlight, thus being commercially implemented in paints and coatings like Green Envirotec for coating interior walls. This chapter covers the use of immobilized enzymes on nanoparticles for air purification. Further, it also emphasizes the utility of combo-technology for commercial implementation.

4.2 Air pollution: an overview

Air pollutants are categorized as biological particles (molds, viruses, parasites, bacteria, spores, animal dander, pollen, etc.), nonbiological particles (smoke, dust, heavy metals, radioactive isotopes, etc.) and gases (fumes from things like adhesives, petroleum products, pesticides, paint, and cleaning products; radon, carbon monoxide, etc.). Generally, fuel combustion, industrial processes, and airplanes are known to be toxic sources of air pollution. A long list of toxic pollutants is found in the atmosphere. Atmospheric air in the troposphere has been analyzed using gas chromatography and mass spectrometry by a team of scientists near the Arizona-Mexico border. They have identified over 586 chemicals containing toxic pesticides like diazinon, chlorpyrifos, and DDT (dichlorodiphenyltrichloroethane), with phthalates at the highest levels. Further, they pointed out that over 120 chemicals could not even be identified [11]. Commonly found pollutants as identified were: molds, bioaerosols, combustion by-products (PAH, CO, CO_2, NO_x), tobacco smoke, formaldehyde, arsenic, volatile organic compounds (VOCs), asbestos, heavy metals (lead, mercury, cadmium, chromium, etc.), and radon (radioactive gas coming from uranium).

Air found outside the home is toxic in addition to indoor air, which is often ignored. Indoor air pollution can come from a variety of sources, like furniture, cabinets, and materials being used for building construction, etc. Indoor air is usually rich in VOCs, released from paints, paint strippers and solvents, wood preservatives, cleaners and disinfectants, moth repellents, air fresheners and aerosol sprays of various kinds, stored fuels, car products, pesticides, copier and printer fluids, dry-cleaned clothing, graphic and craft materials, building materials, correction fluid, hobby supplies, wood glue, permanent markers, pressed wood products made with MDF (medium-density fiberboard), and household cleaning products that have both short- and long-term health effects. Thus we are surrounded by toxic air pollutants whether indoors or outdoors. This has become a more serious problem and is getting worse every day, with the introduction of more and more classes of air pollutants with technology updates in various sectors. Air pollutants have been labeled as air toxics or criteria pollutants based on certain parameters. Air toxics are the category of air pollutants usually present in air at relatively low concentrations. However, they are very toxic at even low concentrations, with the additional property of long-lasting persistence, due to which they are also called hazardous air pollutants. They are very toxic to human, plant, or animal health. Diseases like the incidence of cancer, birth defects, genetic damage, central nervous system defects, immunodeficiency, and disorders of the respiratory and nervous systems have been reported to be caused by air toxics [12, 13]. Their effects are more deleterious to sensitive members of the community including very young and elderly people. These pollutants include volatile as well as semivolatile organic compounds and heavy metals. Various sources of air toxics are motor vehicle emissions, industrial emissions, products of burning fuel, and materials such as paints and adhesives in new buildings. Further, there is a subclass of air toxics called reactive organic compounds that contribute to ozone formation. They are highly toxic and are a significant contributor toward formation of photochemical smog [14]. Other priority air toxics are formaldehyde, toluene, xylene, and polycyclic aromatic hydrocarbons (PAHs). Criteria or common air pollutants are carbon monoxide, lead, nitrogen dioxide, ozone, particles and sulfur dioxide. They are known to be indicators of air quality used for setting standards relating to health and/or environmental effects. These pollutants are widely distributed and are generally found around the world. Countries usually set parameters for levels of air pollution based on criteria air pollutants using carbon monoxide (CO), lead (Pb), nitrogen dioxide (NO_2), ozone (O_3), particles and sulfur dioxide (SO_2). Young children are highly susceptible to the effects of air pollution:

- They often breathe through their mouths rather than their noses, due to which PM can easily escape filtration by nasal cilia in the upper respiratory tract.
- They inhale a larger amount of air pollutants due to their higher ventilation rate, related to their increased heart and respiratory rates.
- Their immune system is not well matured and therefore is more prone to inflammatory and allergic reactions caused by air pollutants.
- They have higher cumulative risk from air pollutants over their life spans.

Prenatal exposure to air pollution has profound effects on the brain of the developing fetus, with additional risks due to genetic abnormalities leading to increased cancer risk, smaller newborn head size, lower birth weight, developmental delays, and higher risk for childhood asthma.

4.3 Nanotechnology and air remediation

Air remediation techniques using nanotechnology can be categorized into four broad types: nanofiltration, catalysis, adsorption, and using sensors for monitoring. These techniques are described in the following paragraphs (see Fig. 4.3).

4.3.1 Nanofiltration

CNTs have been extensively used for removing toxic gases via filtration. Several reports have been published based on the utility of CNTs for air remediation and they have been found to achieve commendable results. CNTs have been fabricated with polymeric nanocomposite membranes to obtain simpler, much faster, and more easily scalable filtering devices for high gas flux transport. Here, fabrication provides a large surface area for the membrane and also improves its filtering efficiency.

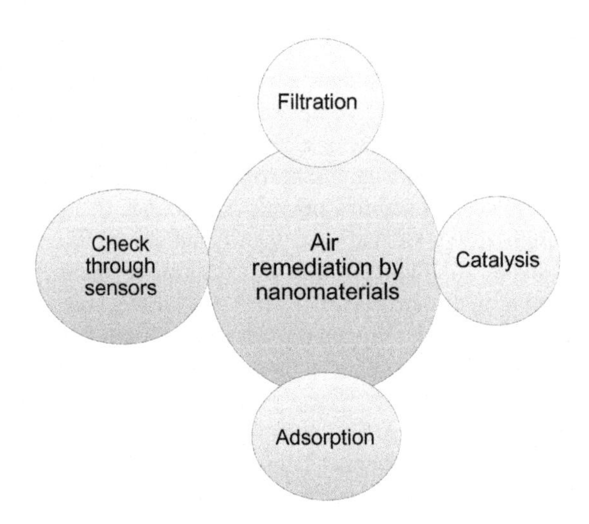

FIG. 4.3

Polluted air can be easily treated by various nanomaterials through various processes.

This membrane was tested for gas mixtures and it was found that the permeation capability was effective in filtering gas mixtures on the basis of their size [15]. It is extremely important that filtration of gases should be such that normal constituents of the atmosphere can easily pass through the nanomaterial-based filtering unit while toxic components are filtered out and retained. However, in the case of CO_2 (a common constituent of the atmosphere), the amount being allowed to pass through the nanofilters should be monitored. Most importantly, filtration should not be affected by variation in the amounts of the gaseous components or by other physical parameters up to a certain limit, such as temperature. A report has been published on adsorption and diffusion of N_2 (largest component of the atmosphere) by single wall carbon nanotubes (SWCNTs). The studies are based on molecular simulation with a range of nanotube diameters (8.61–15.66Å) at temperatures of 100–298K. It was found that nanotubes with diameter corresponding to the size of N_2 were quite effective in its diffusion, regardless of the amount of N_2 in the designed temperature range. There is a distinct organization of layers based on the size of N_2 as confirmed by geometric considerations. Further, the core of the CNTs was more effective in adsorption or diffusion of N_2 than the peripheral regions [16].

A network of CNTs has been effective in selectively capturing VOCs due to the high surface-to-volume ratio and atomistically smooth surfaces. It was found that interfacial adhesion and diffusion are crucial for governing selective gas transport through CNT networks [17]. Net air filters made up of polyurethane nanofiber have been used to capture PM with sizes ranging from 0.3 to 2.5μm. The filtration performance of nanofiber is dependent on fiber diameter and pore size. The fibrous membrane with nanoscale diameter having ultrathin thickness helps in removal of pollutants to large extent, however its performance depends on the type of pollutant being removed. Net filters so prepared are two-dimensional with a diameter of 20nm, in addition to having scaffolded structure with excellent porosity and robust mechanical strength. Polyurethane-based net filters have removal efficiency for $PM_{1-0.5}$ of >99.00% and $PM_{2.5-1}$ of >99.73% at a pressure of 28 Pa. These filters are found to be very effective in air purification, particularly removal of smoke having $PM_{2.5}$, with high efficiency [18]. Hierarchical Ag nanowire has been used to construct a percolation network for filtering air, particularly capturing $PM_{2.5}$ with excellent efficiency. Ag nanowire is transparent and reusable, making it suitable for air remediation applications. It has stronger long-range electrostatic force to capture $PM_{2.5}$ with efficiency >99.99% with less power consumption, and it also has bactericidal activities. [19].

4.3.2 Removal of air pollutants through catalysis

Various nanomaterials are known that can remove toxic air pollutants through catalysis (photocatalysis) and help in purifying air. Titanium dioxide (TiO_2) has been widely used in air remediation through photocatalysis. A highly ordered TiO_2 nanotube (TNT) array was used as an air purifier, particularly in degradation of gaseous acetaldehyde air pollutant, through photocatalysis. It was found that with an increase in the length of the nanotube arrays, the rate of acetaldehyde degradation improved significantly. Further, the porous structure of the TNT array provided efficient diffusion channeling and thus helped in increased acetaldehyde degradation [20]. Nanoparticles of TiO_2 have been found to photocatalytically reduce CO_2 using light-emitting diode (LED) illumination at a wavelength of 388 nm. Further, it was shown that UV illumination is more effective in reducing CO_2, without generating any radical ions [21]. Nanoparticles of TiO_2 have been used in reducing the amount of SO_2 (toxic air pollutant responsible for acid rain) produced during CaO desulfurization on an industrial scale. During optimization, it was found that 8% of TNTs are optimum for good results at a combustion temperature of 850°C. The desulfurization efficiency of TNTs is 87.8% with additional capacity for converting small amounts of SO_2 into SO_3 [22]. The photocatalysis of TNTs under UV-A illumination is used to convert NO and NO_2 into nitrate. The photocatalytic reaction rate of NO is much faster than that of NO_2. The surface of the photocatalyst is extremely important in addition to high crystallinity of TiO_2 for higher efficiency [23]. The organic chemical 1,2-dichlorobenzene found in air can be effectively degraded through photocatalysis by nano-TiO_2 in the wavelength range of 280–380 nm. Nano-TiO_2 in the form of thin film with a pore size of about 100 nm and periodicity of ~150 nm is found to be optimum for photocatalytic activity. The photonic efficiency of nano-TiO_2 is found to be much higher than any other known alternatives [24]. Nano-TiO_2 has been uniformly packed into reactors made of stainless steel and used for the decomposition of benzene and toluene present in air streams. It has been found that the presence of humidity in the air of about 5%–6% is helpful in effective degradation of benzene and toluene. However, a higher level of humidity leads to retardation of decomposition due to competitive adsorption of water molecules onto the active sites of TiO_2 [25]. Indoor air is rich in VOCs, and they are usually ignored unless some toxic health effects are foreseen. Heterogeneous nano-TiO_2 containing anatase (80.7%), brookite

(15.6%), and rutile (3.7%) tricrystalline TiO_2 have been useful in photocatalytically degrading VOCs under UV light illumination. The addition of HNO_3 helps in lowering the temperature for catalysis. Here, heterogeneous nano-TiO_2 can be reused for five cycles without any loss of photocatalysis [26]. Pure nano-TiO_2 with a size of 20 nm can also be used for degradation of VOCs in indoor air under UV irradiation at 253.7 ± 184.9 nm with great efficiency [27].

TiO_2 has been experimented with by mixing it with a number of compounds and observing its catalytic efficiency in degradation of air pollutants. S-doped TiO_2 nanoparticles in the form of $Ti(SO_4)_2$ were used for reducing H_2S gas at a high temperature of 450°C. It was found that a 99.97% reduction was achieved using S-doped TiO_2 [28]. TiO_2 in the form of nanoparticles and nanotubes is supplemented with iron-manganese oxide catalysts in the catalytic reduction of NO. It has been shown that low temperatures up to 230°C can be used for catalysis (much lower with respect to other reports) due to enhanced specific surface area and Brønsted acid sites due to higher Mn^{4+}/Mn^{3+} ratios. The lower temperature has attracted increasing attention due to less energy consumption, thus allowing the technique to be easily employed by industries such as steel plants and glass manufacturing plants. Further, TiO_2 in the form of nanotubes is much more effective than nanoparticles [29]. The mesoporous nanocrystalline C-doped TiO_2 photocatalysts have been prepared and used for degradation of NO under simulated solar light irradiation. The doped TiO_2 prepared by substituting oxygen sites in the TiO_2 lattice by carbon atoms and forming C-Ti-O-C helps in extending the absorption region from UV to visible light [30]. The nanoparticles of TiO_2 were coated on a glass plate, followed by fluorination using a plasma deposition process. The plasma surface–modified TiO_2 nanoparticles were employed as catalysts in the photocatalytic oxidation of *m*-xylene present in the air. The modified TiO_2 was found to be catalytically more active than the unmodified particles. The exact reason for increased photocatalytic activity is not well understood. However, it has been indicated that the plasma surface treatment helps by increasing hydrophobicity of TiO_2, thus providing easier catalyst adsorption of the xylene from the flowing air. Further, fluorinated TiO_2 has a reduced rate of electron-hole recombination, imparting increased photocatalytic activity [31]. TiO_2 nanofibers were doped with cerium (Ce) and used for removal of mercury (Hg^o) present in flue gas, through photocatalytic oxidation. Various parameters have been optimally standardized for the best results, including calcination temperature, cerium dopant content, and illumination conditions. It was found that doping with 0.3 wt% of Ce with a calcination temperature of 400°C led to the highest removal efficiency, 91% for Hg^o removal under UV irradiation. Doping with Ce enhanced the catalytic oxidation by TiO_2 nanofibers, due to the coexistence of Ce^{3+} and Ce^{4+} leading to efficient oxidation of Hg^o present in flue gas. Thus doped TiO_2 nanofibers are helpful in significant suppression of Hg^o emission from flue gas [32]. Nanosized TiO_2/V_2O_5 has also been found useful in removal of elemental mercury (Hg^o) at 300°C. Higher content of V_2O_5 has significantly increased the removal efficiency of elemental mercury, due to the higher amount of vandate species acting as catalyst and lowering temperature from 400°C to 300°C [33]. Nanosized TiO_2/V_2O_5 has also been used in thermal decomposition of 1,2-dichlorobenzene on V_2O_5/TiO_2 nanoparticles. The process is assisted by the presence of vanadium oxytripropoxide and titanium tetraisopropoxide with catalytic reaction temperature at 200°C, leading to 95% of decomposition of gaseous 1,2-dichlorobenzene [34]. Nanosized TiO_2 along with SiO_2 has been immobilized on silica gel and packed into a cylinder-shaped fluidized reactor and used for photocatalytic oxidation of gaseous carbonyl compounds like propionaldehyde, acetone, acetaldehyde, and formaldehyde under UV irradiation. Factors like gas flow rate, relative humidity, concentration of oxygen, and illumination time have been optimized to obtain an excellent rate of photocatalysis. It was found that the rate of degradation of carbonyl compounds followed a decreasing order, as follows: propionaldehyde, acetone, acetaldehyde, and formaldehyde [35]. A novel combination involving TiO_2 nanoparticles and carbonized moso bamboo (*Phyllostachys pubescens*) has been used for removal of toxic pollutants like benzene and toluene from air. Carbonized moso bamboo and TiO_2 were mixed in a 1:1 ratio, leading to effective removal efficiencies of benzene and toluene. Here the presence of moso bamboo helps in enhanced hydrophobic-hydrophobic interaction, a crucial factor in removal of benzene and toluene [36]. The C-N-S-tridoped titanium dioxide (TiO_2) nanocrystals were prepared in the presence of a biomolecule L-cysteine. This prepared novel tridoped nanocrystal has been used for removal of NO under simulated solar light irradiation with excellent removal efficiency with respect to undoped TiO_2 nanocrystals [37]. Small photocatalytic devices based on zirconia-TiO_2 have been used to remove ethylene from closed plant growth units flown in space. Photocatalysis was used for pollutant removal in the presence of low humidity with very high temperatures, for complete oxidation of ethylene. Further, it was found that platinizing zirconia-TiO_2 was useful for improved ethylene oxidation [38]. Synthesized multiwall carbon nanotube (MWNT)-titania (TiO_2) composites have been used for cleaning benzene, toluene, ethyl benzene, and *o*-xylene (usually present in indoor conditions) through photocatalysis. The composite has a removal efficiency much higher than stand-alone TiO_2. This composite was less affected by the presence of humidity; input concentration within a simulated indoor air quality was in the range 0.1–1.0 ppm. The pollutant degradation efficiency was crucially dependent on flow rate; as the flow rate was decreased from 4.0 to 1.0 L min^{-1}, the average efficiency has been increased to 95%–100% [39]. The gaseous benzene can be photocatalytically treated under visible light irradiation by using electrospun $CNTs/TiO_2$ nanofibers. The CNT/TiO_2 nanofibers are further fabricated by electrospinning CNT/poly(vinyl pyrrolidone) (PVP) solution followed by the removal of PVP by calcination at 450°C. The photocatalytic performance of the CNT/TiO_2 nanofibers was found to have a degradation efficiency of 58% for 100 ppm gaseous benzene under visible light irradiation. The larger surface area and lower bandgap energy of CNT/TiO_2 nanofibers were useful for their strong adsorption ability and greater visible light adsorption, thus an effective degradation capability [40].

VOCs are most commonly found in indoor conditions and are mostly ignored unless they produce some serious toxicological health effects. They can easily be degraded by doped nanoparticles, with efficacy. A photoreactor containing TiO_2 and

Pt/TiO_2 photocatalysts has been immobilized onto a quartz support having a total surface area of $4.0 \pm 0.3\,m^2\,g^{-1}$ for the photodegradation of VOCs in situ. A 70% photocatalytic activity was found in the case of toluene (160 ppm) laden air within the first 5 min of UV illumination. This set-up can work continuously for 18 h without disruption. Further, photocatalytic life can be doubled by platinization. The presence of moisture has been found to retard degradation by several folds due to the accumulation of intermediates, which obstructs photocatalysis [41]. The photocatalytic degradation of VOCs through La^{3+} doped TiO_2 coated onto porous nickel mesh has led to a degradation efficiency of VOCs of 86%–94% depending on the content of La^{3+} ranging from 1.5% to 2%. Factors like wind speed and changing wavelength within 254–365 nm have less effect on degradation efficacies [42]; 1.2% La^{3+}doped TiO_2 nanotubes (1.2% La^{3+} TNTs) at 254 nm UV have been used for photocatalytic degradation of gaseous ethylbenzene with degradation efficiency of >80% [43]. It has been further shown that the 0.7% Ln^{3+}-TiO_2 catalysts have the highest adsorption ability while 1.2% Ln^{3+}-TiO_2 catalysts achieved the highest photocatalytic activity [44]. The degradation of VOCs in the range of 1–10 ppm using a photoreactor packed with TiO_2/quartz in the presence of visible light has been carried out cost effectively with a very long life for catalysis. The operational parameters like concentration of pollutant, temperature, and retention time of processing have been optimized, leading to a degradation efficiency of 77.4% at temperatures from 15°C to 35°C, with retention time ranging from 2 to 8.2 s [45]. Clay-supported TiO_2 photocatalysts packed into a benchtop flow reactor have also been used for VOC degradation via photocatalysis. Two composite materials have been used in comparative studies, hectorite-TiO_2 and kaolinite-TiO_2, for degradation of VOCs by using toluene at 170 ppbv as model pollutant at 365 and 254 nm, respectively, in the presence of dry or humid air with relative humidities (RH) of 10%, 33%, and 66%. It was found that hecto-TiO_2 is more efficient than Kao-TiO_2 under UVA irradiation at 365 nm, with reaction rate peaked at 10% RH, leading to effective degradation efficiency [46].

The other metal oxides have also been used as catalysts for degradation of toxic air pollutants. A nanostructured spinel-type $CoCr_2O_4$ catalyst was deposited over a SiC wall flow trap for removal of diesel PM. The catalyst could be regenerated by high-temperature treatment at 550°C. It was best suited for collecting soot agglomerates as well as $PM_{0.1}$ [47]. A nano-NiO/gamma-Al_2O_3 catalyst has been used for tar removal, produced during biomass gasification/pyrolysis. The prepared nanoparticles are spherical in shape with sizes ranging from 12 to 18 nm. These catalysts are packed into a fixed-bed reactor having a tar removal efficiency of 99% at 800°C, with tremendous decrease in the amount of CO_2 and CH_4; however, there is no effect on H_2 and CO [48]. The monolithic nanosized catalyst Pt-Au/ZnO/Al_2O_3 has been used to degrade toluene present in air. The toluene degradation is effective with a lower concentration of toluene at high flow rate, while it decreases drastically with an increase in toluene concentrations [49]. Nanosized α-MnO_2 in the form of rod/wire/tubular and flower-like spherical nanosized Mn_2O_3 has been used for toluene degradation and also compared in terms of their efficacies. The surface area of α-MnO_2 nanorods/nanowires/nanotubes is $45–83\,m^2\,g^{-1}$ while that of flower-like spherical Mn_2O_3 is $162\,m^2\,g^{-1}$. It was found that the order of toluene degradation decreases in the following manner: rod-like α-MnO_2 > tube-like α-MnO_2 > flower-like Mn_2O_3 > wire-like α-MnO_2. The best-performing rod-like α-MnO_2 nanocatalyst can catalyze the total toluene oxidation at 225°C at a space velocity of 20,000 $mL\,g^{-1}\,h^{-1}$ [50]. The monolithic lawn-like CuO-based nanorod array was used for removal of diesel soot produced during combustion at a large scale. It exhibited excellent catalytic activity and was found to be more effective than any known traditional catalysts [51]. Various bimetallic nanostructures like Pd-Ag, Pt-Ag, and Au-Ag have also been used for catalytic CO conversion. Pure metallic nanostructures of Pd, Pt, Au, and Ag have also been used for CO conversion for comparative analyses. It has been shown that Pt has the highest catalytic activity, followed by Pd > Au > Ag. Further, bimetallic nanostructures have higher activities than pure Ag nanostructures [52]. The CNTs have been prepared with the help of Cu/Al_2O_3, Co/Al_2O_3, Fe/Al_2O_3, and Ni/Al_2O_3 catalysts and used for treatment of flue gas containing BTEX, PAHs, SO_2, NO, and CO. The synthesized CNTs from Co/Al_2O_3, Fe/Al_2O_3, and Ni/Al_2O_3 catalysts have a diameter of 20 nm, whereas those synthesized from Cu/Al_2O_3 have a diameter of 50 nm. Pilot-scale test results found that the Co/Al_2O_3 based CNT catalyst is most effective in removal of air pollutants at 250°C [53].

4.3.3 Adsorption of toxic gases

Carbon-based nanoparticles like CNTs, activated carbons, fullerene, and graphene are most commonly exploited for adsorption of toxic gases present in the air. Novel carbon-based adsorbent, 3-amino-propylsilica gel-multiwalled carbon nanotubes (APSG-MW) have been prepared with a surface area of $98\,m^2\,g^{-1}$ and a particle size ranging between 60 and 80 mesh with an average size of 215.0 μm for adsorption of VOCs. This adsorbent is superior to MWCNTs due to its excellent air permeability and was found to be very useful for purifying indoor air [54]. Horn-shaped CNTs have been used for adsorption of various harmful air pollutants like CO_2 and CO at low temperatures. The adsorption capacity for CO_2 is $45\,cm^3\,g^{-1}$ and CO is $17\,cm^3\,g^{-1}$ at 303 K and 850 mm Hg pressure [55]. Mercury has been used for various industrial processes, but it is one of the most toxic elements known. During various processes, mercury is vaporized and enters the atmosphere. The adsorbent based on SWCNTs, highly porous with a very large surface area and light mass density, is useful for mercury adsorption from the air. For 80 mg SWCNT, a maximum adsorption capacity of $0.5\,mg\,g^{-1}$ was achieved within 10 min at 250°C. Further, retention time of mercury on SWCNTs is 3 weeks, and they can be reused for over 30 cycles without any performance loss [56]. Graphene nanosheets have been found to be very useful in adsorption of many toxic air pollutants, in addition to their application in many technological fields, such as energy storage materials, supercapacitors, resonators, quantum dots, solar cells, electronics, and sensors [57]. Magnetic graphene composites have been synthesized and used as an adsorbent for nitropolycyclic aromatic hydrocarbons having $PM_{2.5}$ [58]. SWCNTs have been oxidized by using sodium hypochlorite (NaOCl) and found to be effective adsorbents for isopropyl alcohol (IPA) vapors present in the air stream. The adsorption capacity decreases with increasing temperature and relative humidity. Comparative analyses of IPA adsorption between oxidized SWCNTs and granular activated carbon have revealed that the former is better due to repeated availability of IPA adsorption during 15 cycles of operation [59]. A light membrane composed of SWNTs has been used to encapsulate tetrafluoromethane at 300 K, 1 bar. It was found that the pore size of nanotubes plays a crucial role in determining the efficiency of tetrafluoromethane encapsulation. They are better adsorbents, with efficiency of adsorbing 2.4 mol kg^{-1} of CF_4, than other adsorbents such as activated carbons and zeolites. The best-known substitute, the activated carbon Carbosieve G molecular sieve, can only adsorb 1.7 mol kg^{-1} of CF_4 at 300 K and 1 bar [60]. Two walled MWCNTs have been found to be useful in the adsorption of VOCs, *n*-hexane, benzene, trichloroethylene, and acetone with excellent efficiency [61]. A novel adsorbent APSG-MW (3-amino-propylsilica gel-multiwalled carbon nanotubes) has been prepared by chemical bonding of MWCNTs on silica gel. The surface area of APSG-MW is 98 $m^2 g^{-1}$ with particle size ranging from 60 to 80 mesh having an average size of 215.0 μm. This adsorbent is used for adsorption for VOCs present in indoor air and it has very good stability during storage [54]. Activated carbon nanofibers (ACNFs) having a surface area of 1403 $m^2 g^{-1}$ with large micropore volume of 0.505 $cm^3 g^{-1}$ and narrow average pore diameter of 6.0 Å have been prepared. They are used for toluene adsorption, with adsorption capacity of 65 g-toluene/100 g-ACNF when oxygen to carbon ratio is 1.8% [62]. Carbon nanocages (CNCs) have been synthesized with side length of 220–350 nm and wall thickness of 10–15 nm. They are used as adsorbents for toxic gases like *p*-dihydroxybenzene, *m*-dihydroxybenzene, *o*-dihydroxybenzene, phenol, *m*-cresol, and *o*-cresol as well as PM released during cigarette smoking. The adsorption capability of CNCs has been found to be for phenolic compounds as follows: *p*-dihydroxybenzene (57.31%), *m*-dihydroxybenzene (62.25%), *o*-dihydroxybenzene (65.58%), phenol (75.95%), *m*-cresol (54.34%) and *o*-cresol (59.43%) [63].

TiO_2 nanotubes (TNTs) have been modified with four kinds of amines: monoethanolamine (MEA), ethylenediamine (EDA), triethylenetetramine (TETA) and tetraethylenepentamine (TEPA), and used for adsorbing excess CO_2 (greenhouse gas) present in air. The prepared TNTs are packed into a column and then air is passed through the column for CO_2 adsorption. Here it has been found that TEPA is best suited for providing excellent adsorption capacity for CO_2. TNTs can absorb CO_2 at 30°C with recyclability of five cycles without any loss of performance [64]. A composite membrane made of cellulose acetate impregnated with TiO_2 (CA-TiO_2) was used for adsorption of CO_2 due to limitation by pore size of the membrane. The adsorption kinetics studies on pure and composite membranes have helped in understanding diffusion and solubility of CO_2 through membranes as well as also revealed that the blended CA-TiO_2 membrane adsorbed more quantity of CO_2 gas than the pure CA membrane [65]. Functionlized meso-silica nanoparticles have been prepared to use as adsorbents for formaldehyde (H_2CO) vapor present in contaminated air, most frequently in indoor air. The best part of these nanoparticles is that their synthesis does not produce any pollutant responsible for causing secondary pollution like air, water or soil pollution. The adsorption efficiency so obtained is much higher than any other reported source for H_2CO adsorption [66]. Aminolyted double walled silanes nanotubes (DWSNT) have been prepared to use as effective adsorbent for CO_2 capture. The presence of amino- groups helps in effective CO_2 adsorption in the temperature range 25–100°C. However, the highest efficiency has been found at 100°C [67].

SO_2 has been a great nuisance to the environment, with its effect on generation of secondary pollutants making it more toxic to the atmosphere. Nanoparticles have been very useful in this respect, in the adsorption of SO_2 present in contaminated air. Nanoparticles ranging from 3 to 4 nm provide highly dispersed adsorption sites for SO_2 molecules with adsorption capacity of about 80% while larger nanoparticles block the ultramicropores, which decreases the accessibility of active sites for SO_2 adsorption. Dynamic adsorption has been much more effective on prehumidified materials than dry, with adsorption capacity increased 250% [68]. Nanoporous supports like SBA-15PL silica (platelet morphology) and ethanol extracted pore expanded MC-41 (PME) composite has been modified with polyamines [poly (propyleneimine) with second (G2) and third (G3) generation dendrimers and polyethyleneimine (PEI)]. These nanoporous materials are used for the selective removal of SO_2. It has been shown that PME composite modified with PEI and G3 has significantly higher adsorption capacity for SO_2 which is 4.68 and 4.34 $mmol\,g^{-1}$, respectively, than any combination modification. Further, the working SO_2 adsorption capacity has been found to decrease with increasing temperature while increase with increase in humidity in the gas mixture. However, these nanoporous support cannot be used for several cycles due to decrease in SO_2 adsorption with increasing number of adsorption cycles, due to evaporation of impregnated polyamines [69]. Mono-dispersed Fe_3O_4 nanoparticles have used for adsorption of various gases like benzene, toluene, ethylbenzene and *m*-xylene (BTEX), in addition to SO_2. The adsorption for BTEX is highest with efficiency of $95 \pm 2\%$, followed by SO_2 ($89 \pm 2\%$) and toluene benzene, ethylbenzene ($75 \pm 1\%$, $59 \pm 6\%$, $55 \pm 2\%$) under dry conditions. It has been observed that the adsorption efficiencies decreased as a function of the increase in relative humidity (RH). These nanoparticles can be reused several times without any loss of efficiency [70].

4.3.4 Nanotechnology keeps a check on air pollution through sensors

The emissions of CO_2 and CO from industries and combustion fuels like coal, oil, hydrocarbon, and natural gases have been exponentially increasing day by day, responsible for a major amount of air pollution and climate change. N-type one-dimensional fabricated semiconducting oxides like ZnO-SnO_2 and SnO_2 nanoparticles have been used as a gas sensor, particularly for H_2 and CO. ZnO-SnO_2 nanoparticles are found to be better sensors due to their size and pronounced electron transfer between compound nanostructures, as well as the presence of heterojunctions between ZnO and SnO_2 which provide additional reaction centers [71]. Another ZnO-based CO sensor has been prepared with its operating temperature from 110°C to 180°C and concentration of CO ranging from 100 to 1000 ppm. Obtained sensitivity and the response time of this sensor have presented good performance with increasing operating temperatures and CO concentrations. It has been recommended for large-scale production due to its low cost and excellent sensitivity [72]. A novel sensor based on LaOCl-coating onto ZnO nanowires (NWs) has been prepared for the detection of CO_2 (250–4000 ppm) and CO (10–200 ppm) gases. Here a LaOCl coating helps in increasing sensing performance for the gases, more effectively for CO_2 than CO. Further, the LaOCl coating shortens the response time due to extension of the electron depletion layer, which is due to formation of p-LaOCl/n-ZnO junctions on the surfaces of the ZnO NWs [73]. A zinc oxide (ZnO)/multiwalled carbon nanotubes (MWCNTs) composite-based sensor has been synthesized having improved CO sensing properties with respect to other known CO sensors. The nanocomposite so prepared contains 30 mg ZnO dispersed on 20 mg MWCNTs, demonstrating excellent sensitivity and fast response time ranging from 8 to 23 s for CO with concentrations from 40 to 200 ppm at room temperature [74]. Crystalline nanowires based on $Cd(OH)_2/CdCO_3$ nanowires with lengths in the range from 0.3 to several microns and 5–30 nm in diameter have been prepared. Gas-sensing properties of this sensor for CO_2 have been found to be in the concentration range of 0.2–5 v/v% (2000–50,000 ppm) with suitable performance to be implemented for commercial applications [75].

MoS_2 nanosheets were used for sensing NO_2 present in air at room temperature. However, it was found that NO_2 sensing is more accurate in an inert atmosphere while the performance reduces drastically with oxygen in the air. Therefore, a nanohybrid of SnO_2 nanocrystal (NC) has been decorated with a MoS_2 nanosheet to provide air stability in the presence of oxygen, for sensing NO_2 at room temperature. Hybridization has helped in providing p-type dopants for MoS_2 by SnO_2 NC as well as p-type channels in the MoS_2 nanosheets. This hybrid sensor can work both in dry and humid air without any performance loss and has high sensitivity, excellent selectivity, and repeatability for NO_2 [76]. A novel amperometric type based on a nano-structured CuO sensing electrode and with $La_{10}Si_5NbO_{27.5}$ (LSNO) as the electrolyte has been used as a NO_2 sensor. CuO nanoparticles have a diameter range of 200–500 nm and they are homogeneously dispersed on the LSNO backbone in a porous layer. The response current has been linear for NO_2 with concentrations ranging from 25 to 500 ppm at 600–800°C with sensitivity of 297 nA ppm^{-1} at 800°C [77]. Three-dimensional porous nitrogen-doped nickel oxide (NiO) nanosheets, having excellent p-type semiconducting properties, have been used as an NO_2 sensor. Here, nitrogen doping is used to enhance responsivity and sensitivity of the sensor [78]. A NO_x sensor has been developed that can work at room temperature; it is based on nanocomposite thin film made of cross-linked polyaniline [CLPANI, derived from polyaniline and aniline formaldehyde condensate (AFC)] and fabricated with WO_3. The sensor has an excellent shelf life with response and recovery time of 30 s and 11 min, respectively [79].

Various nanoparticle-based sensors have been developed for indoor air contaminants as discussed. Porous SnO_2 nanospheres with diameter ranging from 90 to 150 nm having high surface areas were used as sensors for VOCs. Contaminants such as hydrochloric acid and NaClO can be determined by porous SnO_2 nanospheres. This sensor has exhibited superior gas-sensing performance for 2-chloroethanol and formaldehyde vapor at ppb level [80]. The ZnO nanowires are used for the detection of VOCs via an infrared (IR) chemical sensing system. The large surface area ZnO nanowires help in effective adsorption of the examined species (VOCs), responsible for the excellent sensitivity from using IR sensing systems. The ZnO nanowires have exhibited better performance for aromatic-type VOCs than for nonaromatic compounds [81]. Three-dimensional polyacrylonitrile/ZnO nanomaterial with very large surface area has been used as an optical sensor for VOC detection. The addition of a polyacrylonitrile nanofiber template has improved the properties of the ZnO nanomaterial, such as surface strain, roughness, and increased nucleation sites. The improved ZnO nanomaterial has been found to be very effective in sensing VOCs even at room temperature [82]. A novel single-crystalline ZnO nanosheet having a porous structure and fabricated with $ZnS(en)_{0.5}$ (en = ethylenediamine) has been synthesized. Fabrication uses a complex precursor having numerous mesopores with diameter of 26.1 nm uniformly distributed along each nanosheet, giving it high density. These novel nanosheet structures are used as sensors for indoor air contaminants like formaldehyde and ammonia. This sensor has exhibited highly sensitive performance with respect to high gas-sensing responses, with short response and recovery times as well as possessing significant long-term stability [83]. An ultraselective and sensitive sensor based on Cr-doped NiO flower-like hierarchical nanostructures has been developed for the detection of indoor air pollutants like xylene, toluene, benzene, ethanol, and formaldehyde. This sensor has shown high responses toward *o*-xylene and toluene at 5 ppm corresponding to a ratio of resistance of gas to air of 11.61 and 7.81, respectively. Further, there are negligible cross-responses toward gases like benzene, formaldehyde, ethanol, H_2, and CO. Here, Cr doping has been very useful in imparting excellent response and selectivity of the sensor toward indoor air contaminants due to a decrease in number of hole concentrations in NiO and catalytic oxidation of methyl groups present in the various gases [84].

Ag/ZnO nanocomposites supported by polycrystalline Al_2O_3 have been used in the detection of flammable and toxic gases [both reducing (CH_3CH_2OH, CH_3COCH_3) and oxidizing (O_3)] in the temperature range 100–400°C. This sensor has high sensitivity and its response is directly proportional to Ag content, which imparts both catalytic and electronic effects to the sensor. Further, the suitable temperature range is helpful for sensing, in the distinguishing between oxidizing and reducing analytes [85]. A nanomaterial-based sensor has been developed that can detect ozone, particularly in the stratosphere. The $CuAlO_2$ nanostructures, having an average size ranging from 40 to 80 nm, have been synthesized with their shapes resembling pentagon and oval structures, having a specific surface area of 59.8 and 70.8 $m^2 g^{-1}$, respectively. It was found that $CuAlO_2$ with pentagonal morphology exhibited a superior response and a better recovery time than the oval-shaped morphology corresponding to 25 and 39 s; 28 and 69 s, respectively at 200 ppb of ozone, 250°C. Thus, particle morphology plays a crucial role, as it is responsible for particle size, surface area, gas adsorption/desorption, and grain-grain contact [86].

4.4 Combo-technology: an additional boon for air remediation

Nanotechnology has found solutions for toxic air remediation. However, there are many shortcomings in addition to those discussed in Chapter 1, as follows:

- *Nanoparticle synthesis:* In most cases, it was found that synthesis of nanoparticles is not eco-friendly. Nanoparticles are known for providing solutions to environmental problems, but their own synthesis poses yet another environmental problem. Further, they cannot be used in their native states as synthesized in the environment, such as in air, due to posing a number of health problems, as discussed in Chapter 1.
- *Operating conditions:* It has been shown that most of the conditions chosen for optimum operation of nanoparticles are not usually found under normal situations. For example, the very high temperatures require additional energy usage and a well-designed set-up for carrying out air remediation.
- *Laboratory conditions:* Most of the studies are performed under laboratory conditions in small-scale experiments. None of them have operated in large-scale plants, which is usually required if successful air remediation is to be done.
- *Humidity:* The atmosphere is a continuously changing phenomenon, dependent on a number of factors. It is never stable in any given time and space. Most importantly, relative humidity is the most frequent variable factor. Most of the studies have found that successful removal of toxic air pollutants can only take place when there is dry air, which is usually not the case under in situ conditions.
- *Interference by other gases:* It was found that removal of toxic air pollutants through catalysis, photodegradation, adsorption, filtration, etc. was successful when only the amount of the particular pollutant (to be treated) is there, while other gases are present minimally or are entirely absent. This is an ideal condition that doesn't normally exist in the atmosphere, as there are a number of gases present due to many factors at a particular time and space.

Combo-technology can be useful, involving the addition of enzymes to nanoparticles to some extent. Many enzymes are being researched that can act directly on toxic pollutants, but only one enzyme is known so far, i.e., carbonic anhydrase (CA). This enzyme has been shown to be very helpful in removal of greenhouses gases (a major climatic challenge) like CO_2 and CO from the atmosphere. CAs are metalloenzymes that play an important role in the biomimetic CO_2 capture process (CCP) through catalysis. CAs are among the fastest known enzymes, with k_{cat} values of up to $10^6\,s^{-1}$ and they have excellent specificity for CO_2. Thermostable CAs are more useful in CCP technology due to their ability to undergo catalysis in extreme physicochemical conditions. CA immobilization onto nanoparticles has been chosen as the best way of utilizing enzymes on an industrial scale for CCP technology, due to its enhanced stability, higher reusability, and its capability of undergoing catalysis even in extreme physicochemical conditions, with respect to soluble CA. Further, this reduces the total cost of the process by several hundred folds. Various constraining factors with respect to nanoparticle usage in air remediation, like operating conditions, laboratory conditions, humidity, and interference by other gases, can easily be rectified due to the high specificity and catalysis under a broad range of physicochemical conditions. Toxicity due to nanoparticle can be minimized to some extent by using eco-friendly nanoparticles, but that does not solves the overall problem. Supplementing with enzyme technology, viz. in the present case CA has been immobilized onto various nanoparticles and implemented for air remediation as given in in following section.

A human carbonic anhydrase (HCA) has been immobilized onto gold nanoparticles assembled over amine/thiol-functionalized mesoporous SBA-15, forming HCA/Au/APTES/SBA-15 and HCA/Au/MPTES/SBA-15, respectively. The performance of the immobilized enzyme has been studied by using hydrolysis of para-nitrophenyl acetate (*p*-NPA). The kinetic parameters obtained for HCA/Au/APTES/SBA-15 and HCA/Au/MPTES/SBA-15 are: K_m (22.35 and 27.75 mM), and k_{cat}/K_m (1514.09 and 1612.25 $M^{-1} s^{-1}$), respectively. These immobilized enzymes are able to undergo hydration of CO_2 and its subsequent precipitation as $CaCO_3$. The amount of $CaCO_3$ precipitated has been found to be almost the same in the case of soluble and immobilized enzymes. However, storage stability and reusability of immobilized enzymes, as well as their adaptability to any reactor design, make them more suitable for CO_2 sequestration. Here, storage stability and reusability of HCA/Au/MPTES/SBA-15 is much higher than HCA/Au/APTES/SBA-15, as the former retained its activity even after 20 days storage at 25°C with 20 recycling runs. Both immobilized enzymes are recommended for industrial-scale usage in CO_2 sequestration [87].

Bovine carbonic anhydrase isoform II (BCA II) has been immobilized onto KIT-6 nanoporous silica nanoparticles having a diameter of <100 nm. Structural studies by circular dichroism and fluorescence spectroscopy have found that immobilized enzyme is highly stable even at very high concentrations of denaturants. The melting temperatures (T_m) of soluble and immobilized BCA II have been found to be 64.7°C and 71.9°C, respectively. Further immobilized enzyme has pronounced stability against pH and thermal deactivation. Immobilized BCA II has been strongly recommended for industrial-scale applications, particularly in CO_2 capturing of large amounts [88]. A large amount of CA has been obtained through DNA recombinant technology by using CA DNA from thermophilic bacterium *Sulfurihydrogenibium yellowstonense* and expressed in *Escherichia coli* and labeled as SspCA (α-CA). This enzyme has been immobilized onto the surface of magnetic Fe_3O_4 nanoparticles (MNP) through carbodiimide. Here, various growth parameters for culturing *E. coli* leading to maximum enzyme production have been intensely optimized. Both enzyme stability and storage time have improved significantly during immobilization and most importantly immobilized enzyme can be easily recovered once it has undergone catalysis through application of a magnet or electromagnetic field [89]. A bioreactor based on continuous biocatalytic hydration of CO_2 has been performed by using CA from *Rhodobacter sphaeroides* immobilized onto electrospun polystyrene/poly(styrene-*co*-maleic anhydride) (PS/PSMA) nanofibers in the form of cross-linked enzyme aggregates (CLEA). The CA-CLEA has a storage stability of 60 days at 4°C with 60 reuses, without any performance loss. A large amount of enzyme can be produced by enhanced cell growth of *R. sphaeroides* with the addition of organic substances like carotenoid, bacteriochlorophyll, porphyrin, and coenzyme Q10 in the growth medium. Immobilized enzyme has been produced in an environmentally friendly manner, which has further added momentum for its greater usage in large-scale plants [90]. The conversion of CO_2 into biocarbonate was performed in a biomimetic nanoconfiguration by using immobilized CA onto carboxylic acid group-functionalized mesoporous silica (HOOC-FMS). The enzyme immobilization led to a slight change in conformational structure of the enzyme. This can easily be compensated for with a large amount of enzyme loading, which provides higher enzymatic activity as well as higher immobilization efficiency of >60%. This work has provided a novel platform for converting CO_2 into biocarbonate, which can easily be incorporated into other biosynthetic processes [91].

Immobilized CA can be used for air remediation in a pilot plant set-up in the form of biofilters. The treatment of polluted air by biological means using immobilized enzymes would be helpful for industries generating enormous amounts of pollutants released directly into the air without any additional toxicological side effects. Further, it is cost effective and environmentally friendly. Enzyme-based waste air is an effectual substitute with less energy input and it is easily adaptable by both small-scale and large-scale industries without much expertise. Additionally, it works in the temperature range 30–40°C without generation of any toxic waste. In various set-ups, enzyme-expressing microbes have been immobilized onto suitable support and packed into column-like structures with suitable optimization of growth media conditions. For example, immobilized pure culture of *Burkholderia cepacia* PR123 and *Pseudomonas putida* have been found to constitutively express ortho-monooxygenase for the biodegradation of VOCs [92]. Sometimes, DNA of the enzyme of interest is artificially introduced into the bacteria through recombinant technology and used for degradation of air pollutants. Usually, these microbes are immobilized onto inert supports like perlite, polypropylene Pall rings, structured plastic, polyurethane foam, lava rocks, etc. They usually require a large amount of pollutants for injection so that enzymes can be induced for expression by microbes. The time required for pollutant degradation is very large, viz. microbial wet biomass (3.1–9.8 $kg\,m^{-3}\,day^{-1}$) requires about 2–3 weeks for complete degradation of nitrobenzene with inlet concentration of 80 $mg\,m^{-3}$ [93]. It was also reported that production of secondary pollutants might take place due to chemical reactivity with various factors present in growth media. The other greatest challenge is the clogging of biofilters due to biomass accumulation, which drastically destroys the pollutant-degrading capabilities of immobilized microbes. Biofilm generally develops on the surface of microbes, due to which growth of the microbe-producing enzyme is suppressed and completely stopped as the biofilm thickens. Essential factors, particularly mineral nutrients and oxygen, are blocked due to biofilm formation as indicated by scanning confocal laser microscopy, computed axial tomography (CAT), and X-ray studies. Further, biofilm formation obstructs microbial growth responsible for producing the enzyme for pollutant degradation. Several remedies have been adopted, such as addition of higher concentrations of NaCl (above optimum concentration) or lower concentrations of potassium or nitrogen (below optimum concentration), or usage of preying organisms like protozoa to regulate biomass accumulation. There should be regular removal of excess biomass during biofilter functioning. Further, there should be regular cleaning by using 0.1 M NaOH solutions to obtain stable biomass content, once every 2 weeks. Thus, the complete process has proven to be tedious and requires a great deal of expertise to solve various problems generated during biofilter operation [94].

Immobilized enzyme-based biofilters have been found to be very useful in air remediation without generating any problems as discussed in the preceding section (Table 4.1). Nanoparticle-based enzyme immobilization has been reported to impart much stability toward extreme physicochemical conditions, storage stability, higher number of reuses, and effective catalytic properties. Due to their higher specificity, very large amounts of substrate (pollutant) are not required for action. Further, the reaction medium is very simple, containing a proper buffer at specific pH and a few additional factors as peer enzyme requirements with operational temperature from 30°C to 40°C. Polluted air is introduced into the column packed with immobilized enzyme followed by enzyme action and release of remediated air. Presently, CA has only been immobilized onto nanoparticles and used for removal of CO and CO_2 from toxic air. Design of column and immobilized enzyme packing are properly optimized for excellent performance with air flow generally recommended in the downward direction for efficacy [95]. There is least adsorption of pollutants onto supporting bed due to its inert nature. Nanoparticle-based enzyme columns are useful in performing air filtration, as they also remove various other pollutants nonspecifically due to size constraints, as presented by the immobilized enzyme onto nanoparticles in addition to catalytic pollutant degradation. Due to excellent enzyme specificity, pollutants are degraded as soon as they are introduced into the column, thus saving time. Maintenance cost is much less than any set-up known so far, requiring optimization of a few factors like operating pH and temperature, reactor stability, and proper maintenance of various additives (required during enzymic reactions).

Table 4.1 Recalcitrant Air Pollutants Can Be Effectively Degraded by Immobilized Microorganisms Onto Biofilter Containing Various Enzymes

Substrate	Microbe	Degradation Product
Methanol	*Pseudomonas*	Water, carbon dioxide
Dimethylamines	*P. aminovorans*	Methylamine and formaldehyde
Phenol	*P. putida*	Acetaldehyde and pyruvate
Benzaldehyde	*Acetobacter ascendens*	Benzyl alcohol and benzoic acid
Aniline	*Nocardia* sp. and *Pseudomonas*	Pyrocatechol
Indole	*Chromobacterium*	Pyrocatechol
Camphor	*P. putida*	Lactonic acid

The work is in progress to explore more and more enzymes for degradation of air pollutants with suitable immobilizing nanoparticles, which impart excellent stability without compromising any reduction in catalytic properties of enzymes. This will need more extended application of the modern tools of nanotechnology and enzyme technology, which will enable the establishment of baseline information for rational reactor design and its operation under optimum conditions. Further, it would also be helpful in scaling up to pilot scale plant for degradation of various recalcitrant pollutants present in the air.

4.5 Future aspects

Nanotechnology is providing solutions to various environmental problems. Nanomaterials have properties that enable chemical reduction, catalysis, adsorption, or even filtration to mitigate air pollutants of concern. Although nanomaterials have beneficial applications, concerns have also risen over potential implications for human health and the environment. Important factors of concern are: bioaccumulation potential, toxicity, worker and community exposure. The remediation of environmental pollution through the applications of nanotechnology should be undertaken with consideration of potential impacts across the entire lifecycle of the nanomaterials from their production, usage, to end-of-life disposition [96, 97]. Materials at nanoscale pose more toxicological concerns, viz. airborne nanoparticles pose serious lung disease risk, as it was observed recently that CNTs have acted like asbestos particles on being introduced into the human body when inhaled in sufficient amount. The poor understanding of the fate and behavior of nanoparticles in humans and the environment have exacerbated concerns, with scarce studies focusing on the effects of nanomaterials on human health and environment. Various international organizations like RCEP (Royal Commission on Environmental Pollution) and EU (European Union) are well acquainted with laboratory testing on various nanomaterials, which suggests that they have properties that could cause concern [98]. This is the massive challenge for scientific authorities: to monitor the consequent impact of the huge volume of diverse nanoparticles being produced and used. There are very few reports that practically correlate between theoretical risks and real monitoring of toxicological effects of nanoparticles on human health and the environment.

Air remediation using nanoparticles has been quite lucrative and has presented various important results. Undoubtedly, nanoparticles have presented a better substitute with respect to other known conventional technologies, in terms of performance and efficacy. However, there are various limitations besides health and environmental effects. Nanoparticle-based air remediation is an energy-consumptive process with a requirement of high operating temperature. Further, the process of air remediation is greatly affected by various additional factors that are practically not possible to remove during air remediation. Therefore, there is an urgent need for additional technology that can complement nanotechnology with removal of any toxic effects while adding more lucrative features to help in giving a complete solution to environmental problems. Combo-technology involving nanotechnology and enzyme technology has created lots of enthusiasm among scientists, to work on and find solutions not only to those issues being raised from nanotechnology but to greater environmental problems. There are very few enzymes that are being explored for air remediation after being immobilized onto nanoparticles. Thus, an urgent need exists to explore more enzymes and immobilize them appropriately onto suitable nanoparticles. Nanoparticles as sensors for various air pollutants have been very successful and quite acceptable due to having no side effects [99]. Therefore, nanotechnology is also very useful in monitoring environmental conditions in addition to helping find solutions leading to the eradication of environmental problems. The time is no longer so far off when there will be solutions to many serious environmental problems through usage of combo-technology, due to its adaptability, acceptability, and accessibility in the commercial markets.

References

[1] Hamra GB, Guha N, Cohen A, Laden F, Raaschou-Nielsen O, Samet JM, Vineis P, Forastiere F, Saldiva P, Yorifuji T, Loomis D. Environ Health Perspect 2014;122:906–11.

[2] Cichowicz R, Wielgosiński G, Fetter W. Environ Monit Assess 2017;189:605. https://doi.org/10.1007/s10661-017-6319-2.

[3] Rajagopalan S, Brook RD. Glob Heart 2012;7:207–13.

[4] Breysse PN, Diette GB, Matsui EC, Butz AM, Hansel NN, McCormack MC. Proc Am Thorac Soc 2010;7:102–6.

[5] Gil L, Adonis M, Cáceres D, Moreno G. Rev Med Chil 1995;123:411–25.

[6] Smith KR, Mehta S, Feuz M. In: Ezzati M, Lopez AD, Rodgers A, Murray CJL, editors. Comparative quantification of health risks: the global burden of disease due to selected risk factors. vol. 2. Geneva: World Health Organization; 2004. p. 1435–93.

[7] Du Y, Xu X, Chu M, Guo Y, Wang J. J Thorac Dis 2016;8:E8–19.

[8] Lodovici M, Bigagli E. J Toxicol 2011;2011:487074. https://doi.org/10.1155/2011/487074.

[9] Songolzadeh M, Soleimani M, Ravanchi MT, Songolzadeh R. Sci World J 2014;2014:828131. https://doi.org/10.1155/2014/828131.

[10] Rashid H-O, Ralph SF. Nanomaterials (Basel) 2017;7:99. https://doi.org/10.3390/nano7050099.

[11] Suh HH, Bahadori T, Vallarino J, Spengler JD. Environ Health Perspect 2000;108:625–33.

[12] Clark-Reyna SE, Grineski SE, Collins TW. Fam Community Health 2016;39:160–8.

[13] Woodruff TJ, Axelrad DA, Caldwell J, Morello-Frosch R, Rosenbaum A. Environ Health Perspect 1998;106:245–51.

[14] Weisel CP. Environ Health Perspect 2002;110:527–37.

[15] Kim S, Jinschek JR, Chen H, Sholl DS, Marand E. Nano Lett 2007;7:2806–11.

[16] Arora G, Wagner NJ, Sandler SI. Langmuir 2004;20:6268–77.

[17] Wan Y, Guan J, Yang X, Zheng Q, Xu Z. Phys Chem Chem Phys 2014;16:14894–8.

[18] Zuo F, Zhang S, Liu H, Fong H, Yin X, Yu J, Ding B. Small 2017;13:https://doi.org/10.1002/smll.201702139.

[19] Jeong S, Cho H, Han S, Won P, Lee H, Hong S, Yeo J, Kwon J, Ko SH. Nano Lett 2017;17:4339–46.

[20] Liu Z, Zhang X, Nishimoto S, Murakami T, Fujishima A. Environ Sci Technol 2008;42:8547–51.
[21] MacFarlane JW, Scott TB. J Hazard Mater 2012;211:247–54.
[22] Wang SQ, Zhao Y, Tan Q, Xu PY. Huan Jing Ke Xue 2008;29:518–24.
[23] Nguyen NH, Bai H. J Environ Sci (China) 2014;26:1180–7.
[24] Ren M, Ravikrishna R, Valsaraj KT. Environ Sci Technol 2006;40:7029–33.
[25] Ku Y, Chen JS, Chen HW. J Air Waste Manage Assoc 2007;57:279–85.
[26] Chen K, Zhu L, Yang K. J Environ Sci (China) 2015;32:89–95.
[27] Jwo CS, Chang H, Kao MJ, Lin CH. J Nanosci Nanotechnol 2007;7:1947–52.
[28] Shahzad N, Hussain ST, Siddiqua A, Baig MA. J Nanosci Nanotechnol 2012;12:5061–5.
[29] Lee T, Liou S, Bai H. J Air Waste Manage Assoc 2017;67:292–305.
[30] Huang Y, Ho W, Lee S, Zhang L, Li G, Yu JC. Langmuir 2008;24:3510–6.
[31] Sumitsawan S, Cho J, Sattler ML, Timmons RB. Environ Sci Technol 2011;45:6970–7.
[32] Wang L, Zhao Y, Zhang J. Chemosphere 2017;185:690–8.
[33] Lee W, Bae GN. Environ Sci Technol 2009;43:1522–7.
[34] Chin S, Jurng J, Lee JH, Moon SJ. Chemosphere 2009;75:1206–9.
[35] Zhang M, An T, Fu J, Sheng G, Wang X, Hu X, Ding X. Chemosphere 2006;64:423–31.
[36] Chuang CS, Wang MK, Ko CH, Ou CC, Wu CH. Bioresour Technol 2008;99:954–8.
[37] Wang Y, Huang Y, Ho W, Zhang L, Zou Z, Lee S. J Hazard Mater 2009;169:77–87.
[38] Tibbitts TW, Cushman KE, Fu X, Anderson MA, Bula RJ. Adv Space Res 1998;22:1443–51.
[39] Jo WK. J Air Waste Manage Assoc 2013;63:963–70.
[40] Wongaree M, Chiarakorn S, Chuangchote S, Sagawa T. Environ Sci Pollut Res Int 2016;23:21395–406.
[41] Young C, Lim TM, Chiang K, Amal R. Water Sci Technol 2004;50:251–6.
[42] Ding Z, Feng XG, Chen XD, Fu DG, Yuan CW. Huan Jing Ke Xue 2006;27:1814–9.
[43] Cheng ZW, Feng L, Chen JM, Yu JM, Jiang YF. J Hazard Mater 2013;254:354–63.
[44] Li FB, Li XZ, Ao CH, Lee SC, Hou MF. Chemosphere 2005;59:787–800.
[45] Tsai CW, Chang CT, Chiou CS, Shie JL, Chang YM. J Air Waste Manage Assoc 2008;58:1266–73.
[46] Kibanova D, Cervini-Silva J, Destaillats H. Environ Sci Technol 2009;43:1500–6.
[47] Cauda E, Hernandez S, Fino D, Saracco G, Specchia V. Environ Sci Technol 2006;40:5532–7.
[48] Li J, Yan R, Xiao B, Liang DT, Du L. Environ Sci Technol 2008;42:6224–9.
[49] Kim KJ, Ahn HG. J Nanosci Nanotechnol 2015;15:6108–11.
[50] Wang F, Dai H, Deng J, Bai G, Ji K, Liu Y. Environ Sci Technol 2012;46:4034–41.
[51] Yu Y, Meng M, Dai F. Nanoscale 2013;5:904–9.
[52] Byeon JH, Kim JW. ACS Appl Mater Interfaces 2014;6:3105–10.
[53] Lu CY, Tseng HH, Wey MY, Chuang KH, Kuo JH. J Environ Manag 2009;90:1884–92.
[54] Wang L, Liu J, Zhao P, Ning Z, Fan H. J Chromatogr A 2010;1217:5741–5.
[55] Sawant SY, Somani RS, Bajaj HC, Sharma SS. J Hazard Mater 2012;227:317–26.
[56] Golbabaei F, Ebrahimi A, Shirkhanloo H, Koohpaei A, Faghihi-Zarandi A. Global J Health Sci 2015;8:273–80.
[57] Yu JG, Yu LY, Yang H, Liu Q, Chen XH, Jiang XY, Chen XQ, Jiao FP. Sci Total Environ 2015;502:70–9.
[58] Zhang J, Zhang L, Li R, Hu D, Ma N, Shuang S, Cai Z, Dong C. Analyst 2015;140:1711–6.
[59] Hsu SC, Lu C. J Air Waste Manage Assoc 2009;59:990–7.
[60] Kowalczyk P, Holyst R. Environ Sci Technol 2008;42:2931–6.

[61] Shih YH, Li MS. J Hazard Mater 2008;154:21–8.
[62] Oh GY, Ju YW, Kim MY, Jung HR, Kim HJ, Lee WJ. Sci Total Environ 2008;393:341–7.
[63] Li G, Yu H, Xu L, Ma Q, Chen C, Hao Q, Qian Y. Nanoscale 2011;3:3251–7.
[64] Song F, Zhao Y, Ding H, Cao Y, Ding J, Bu Y, Zhong Q. Environ Technol 2013;34:1405–10.
[65] Hafeez S, Fan X, Hussain A, Martín CF. J Environ Sci (China) 2015;35:163–71.
[66] Ewlad-Ahmed AM, Morris MA, Patwardhan SV, Gibson LT. Environ Sci Technol 2012;46:13354–60.
[67] Ko YG, Lee HJ, Oh HC, Choi US. J Hazard Mater 2013;250:53–60.
[68] Arcibar-Orozco JA, Rangel-Mendez JR, Bandosz TJ. J Hazard Mater 2013;246:300–9.
[69] Tailor R, Abboud M, Sayari A. Environ Sci Technol 2014;48:2025–34.
[70] Ye CZ, Ariya PA. J Environ Sci (China) 2015;31:164–74.
[71] Chen W, Li Q, Xu L, Zeng W. J Nanosci Nanotechnol 2015;15:1245–52.
[72] Hsiao C-C, Luo L-S. Sensors (Basel) 2014;14:12219–32.
[73] Van Hieu N, Khoang ND, Trung do D, Toan le D, Van Duy N, Hoa ND. J Hazard Mater 2013;244:209–16.
[74] Alharbi ND, Ansari MS, Salah N, Khayyat SA, Khan ZH. J Nanosci Nanotechnol 2016;16:439–47.
[75] Krishnakumar T, Jayaprakash R, Prakash T, Sathyaraj D, Donato N, Licoccia S, Latino M, Stassi A, Neri G. Nanotechnology 2011;22:325501https://doi.org/10.1088/0957-4484/22/32/325501.
[76] Cui S, Wen Z, Huang X, Chang J, Chen J. Small 2015;11:2305–13.
[77] Wang L, Han B, Dai L, Zhou H, Li Y, Wu Y, Zhu J. J Hazard Mater 2013;262:545–53.
[78] Luan VH, Tien HN, Hur SH, Han JH, Lee W. Nanomaterials (Basel) 2017;7:313. https://doi.org/10.3390/nano7100313.
[79] Kaushik A, Khan R, Gupta V, Malhotra BD, Ahmad S, Singh SP. J Nanosci Nanotechnol 2009;9:1792–6.
[80] Li Z, Zhao Q, Fan W, Zhan J. Nanoscale 2011;3:1646–52.
[81] Yang J, Shih YR, Chen IC, Kuo CI, Huang YS. Appl Spectrosc 2005;59:1002–8.
[82] Viter R, Abou Chaaya A, Iatsunskyi I, Nowaczyk G, Kovalevskis K, Erts D, Miele P, Smyntyna V, Bechelany M. Nanotechnology 2015;26:105501https://doi.org/10.1088/0957-4484/26/10/105501.
[83] Liu J, Guo Z, Meng F, Luo T, Li M, Liu J. Nanotechnology 2009;20:125501https://doi.org/10.1088/0957-4484/20/12/125501.
[84] Kim HJ, Yoon JW, Choi KI, Jang HW, Umar A, Lee JH. Nanoscale 2013;5:7066–73.
[85] Simon Q, Barreca D, Gasparotto A, Maccato C, Tondello E, Sada C, Comini E, Devi A, Fischer RA. Nanotechnology 2012;23:025502https://doi.org/10.1088/0957-4484/23/2/025502.
[86] Thirumalairajan S, Mastelaro VR, Escanhoela CA. ACS Appl Mater Interfaces 2014;6:21739–49.
[87] Vinoba M, Lim KS, Lee SH, Jeong SK, Alagar M. Langmuir 2011;27:6227–34.
[88] Khameneh HP, Bolouri TG, Nemati F, Rezvani F, Attar F, Saboury AA, Falahati M. Int J Biol Macromol 2017;99:739–45.
[89] Perfetto R, Del Prete S, Vullo D, Sansone G, Barone CMA, Rossi M, Supuran CT, Capasso C. J Enzyme Inhib Med Chem 2017;32:759–66.
[90] Park JM, Kim M, Lee HJ, Jang A, Min J, Kim YH. Biomacromolecules 2012;13:3780–6.
[91] Yu Y, Chen B, Qi W, Li X, Shin Y, Lei C, Liu J. Microporous Mesoporous Mater 2012;153:166–70.
[92] Weaver VB, Kolter R. J Bacteriol 2004;186:2376–84.

[93] Nishino SF, Spain JC. Appl Environ Microbiol 1993;59:2520–5.
[94] Cabrol L, Malhautier L, Poly F, Lepeuple AS, Fanlo JL. FEMS Microbiol Ecol 2012;79:260–71.
[95] Bose H, Satyanarayana T. Front Microbiol 2017;8:1615. https://doi.org/10.3389/fmicb.2017.01615.
[96] Ray PC, Yu H, Fu PP. J Environ Sci Health C Environ Carcinog Ecotoxicol Rev 2009;27:1–35.
[97] Contado C. Front Chem 2015;3:48. https://doi.org/10.3389/fchem.2015.00048.
[98] Yunus IS, Harwin, Kurniawan A, Adityawarman D, Indarto A. Environ Technol Rev 2012;1:136–48.
[99] Penza M, Spetz AL, Romano-Rodriguez A, Beilstein MM. J Nanotechnol 2017;8:2015–6.

CHAPTER

Assessment of combo-technology based on its sustainability and life cycle implications

5

Alka Dwevedi

Department of Biotechnology, Delhi Technological University, New Delhi, India Swami Shraddhanand College, University of Delhi, New Delhi, India

5.1 Introduction

Environmental remediation has been a great concern, especially during the last two decades. Several strategies have appeared, each with its own pros and cons. Currently, combo-technology involving a combination of nanotechnology and enzyme technology has found solutions to most of the environmental-related issues. However, acceptance of any technology does not only depend on its efficacy and cost but also on other factors, primarily on its sustainability [1]. Further, the process should consume the least possible amount of energy and promote ecological revivification by restoring natural flora and fauna. The social and economical issues should be properly dealt with and be transparent to the public as well to government before being allowed to be released into the marketplace. There are issues with other technologies despite their larger claims, the foremost being their cost of operation and maintenance along with their long-term effects on the environment. This process of achieving standards of remediation, as stipulated by state and federal regulators, can take a long time. Therefore, release of any technology, combo-technology in the present case, should be of that which has the least consumption of energy, which is most efficient in permanent removal of contaminants, and which enhances energy production to help in recovery of the economy as well as restoration of the social structure.

Combo-technology can contribute toward sustainable development if it follows the path of human well-being, in addition to economic support, with aspirations of a clean and healthy environment and contributions toward social development. Collectively, it can be said that fulfilling three parameters would provide the tools for sustainable development: environmental, social, and economic [2]. It would be interesting to study the long term effect of combo-technology in poorer countries, especially with respect to development opportunities due to its efficacy in environmental remediation and energy production. The present technology can be useful

Solutions to Environmental Problems Involving Nanotechnology and Enzyme Technology
https://doi.org/10.1016/B978-0-12-813123-7.00005-0

for environmental remediation and energy production in many parts of the world. The point has also been raised that the diffusion of this technology worldwide will only be accomplished by efficient use of natural and environmental resources, with generation of better employment opportunities. Rich countries can take the risk of using combo-technology due to their stable economic growth, but in the case of developing and underdeveloped countries, long-term sustainability is a must for its adoption [3]. However, due to reduced pressures being imposed by environmental policies, developing and underdeveloped countries are open to any new technology, but new technology will only endure if it is sustainable. It can be said that nature itself imposes restrictions on any new technology. There needs to be a strong regulatory framework for environment protection regardless of GDP (gross domestic product) status of any country. This will help in promoting positive structural effects, thus improving use and management of environmental resources as well as bringing about wider diffusion of environmentally friendly technologies. Further, it has been found that environmental regulations have a strong impact on economic growth over the long term.

The social effect of any new technology is also a parameter determining its long-term success, as it directly defines the distribution of income, both within and across nations, as well as affects the reduction of poverty. Technology that boosts the demand for labor leads to increased employment as well as higher monthly wages. Therefore, policymakers have to take into consideration both environmental and social factors before approving any updated technology [4]. Combo-technology must satisfy all the mentioned factors to actually be a success, despite having many advantages in both environmental remediation and clean energy production.

It is an important aspect of any new technology to be sustainable in terms of economic growth, environmental protection, and social development, as well as being well coordinated with national and international policy frameworks. It has been a challenge for developing and underdeveloped countries to introduce any new technology/product, due to lack of adequate skills, knowledge, and competencies. This factor actually leads to increased costs and thus, despite being eco-friendly and having many advantages over existing technologies, updated technology like this is unable to take a front seat in the market. There are some instances where, despite the new product or the technology being environmentally favorable with beneficial social impacts, it was not able to do much in terms of revenue generation and trade expansion/investments [5]. Therefore, some additional factors must be taken care of in addition to eco-labeling, extended producer responsibility, and green public purchasing. Coordination between national and international policies is an important element. However, factors such as effects on human and animal health or plant life and conservation of natural resources remain constant at all policy levels [6]. There should be compliance between the country developing the technology and the country to which the technology is being transferred. Tariff reductions should be considered based on economic standards of the beneficiary country. Therefore, globalization and sustainable development need to work hand in hand for any technology to succeed on an international platform.

International financial institutions and multilateral development banks are another arm helping with lending policies and practices. Their regulatory policies based on standards of environmental, social, and disclosure policies, with additional obligations regarding adherence to international standards, help beneficiary countries in adopting new technology. In the international marketplace, recent guidelines deal explicitly with disclosure (all the facts including side effects), transparency with respect to social and environmental accountability, employment opportunities, and environmental consequences that may follow the release of any new technology or product in the market. This chapter provides an overview on the sustainability of combo-technology and its efficacy in domestic and international spheres, besides addressing the environmental, social, and economic consequences.

5.2 Concept of sustainability and its management

The concept of sustainability has been known for several decades, but its definition evolved over that period. Here we focus on the basic structure from which the idea of sustainability has come into existence. Further, emphasis is placed on its proper management with respect to any technology being developed and brought to the marketplace for use by humankind. Structural components of sustainability are environmental, social, and business, which are discussed in detail in the following paragraphs.

(a) *Environmental Discourse:* This is the most important component of sustainability, based on the fact that sustainability of any product of technology can only prevail if there is firm correlation with nature [7]. Over 30 years ago, the significant role of sustainability as an environmental factor was realized. In 1972, the UN Conference on Human Environment took place in Stockholm and laid the basis for environmental concerns of sustainability. The basic principle defined it as: "The maintenance, restoration and improved capacity of the Earth to produce vital renewable resources for various applications." This led to the foundation of the UN Environmental Program (UNEP) and a number of national environmental protection agencies. UNEP is known for bringing an environmental component to sustainability, as it put forth that any technology or product can only be the source of ecodevelopment if it is based on higher usage of renewable resources, with effective monitoring of depletion of nonrenewable resources. Further, in 1980, the International Union for Conservation of Nature (IUCN), the World Wildlife Fund (WWF), and UNEP developed the World Conservation Strategy (WCS), which has clearly defined that development can only be sustainable if it involves both improvements in human life and conservation of natural resources. Further, the greatest level of sustainability can only be given to present generations as well as future generations with the proper management of human use of the biosphere [8]. Finally, in 1987, a definition of sustainable development was stated by the World Commission on Environment and Development (WCED) as development that meets the needs of the present without any compromising of coming generations in the meeting of their needs. The RIO EARTH SUMMIT by the UN Conference on Environment and Development (UNCED) in 1992 provided a global action plan for sustainability, with special emphasis on environmental aspects. Later, the Kyoto Conference on Climate Change in 1997 moved for a proactive approach for ultimate success in sustainability by incorporating business advantages from the management of environmental performance. This laid the foundation of Environmental Management Systems (EMSs) based on integration of environmental issues into business culture along with their management processes. This involves a database embracing indicators like energy efficiency, material efficiency, biodiversity, emissions, water consumption, and waste. These indicators are monitored, reported, and summarized to provide complete information on environmental performance to internal and external stakeholders of the firm for effective sustainability.

(b) *The Social Discourse:* The social aspects of sustainability have accompanied the discussion on the environmental aspects. In fact, both components were considered parallel when sustainability was being conceptualized [9]. According to the WCED, the social aspect of sustainability includes social justice, distributive justice, and equality of conditions. According to corporate social responsibility (CSR), this is an essential obligation for business success, to comply with objectives and values of our society beyond focusing on direct economic and technical interest. Social responsibility implies a public posture toward society's economics of human resources, with emphasis on usage of resources considering broad social ends not limited to the interests of private persons or companies [10]. As per the UN, social concerns involve themes such as poverty, health, and discrimination. In fact, the Johannesburg World Summit on Sustainable Development (WSSD) in 2002 primarily focused on the concept of sustainability by considering social and economic development, with less emphasis on environmental issues. However, later environmental issues were given a priority by WSSD. Most significantly, a 20-year follow-up of the RIO EARTH SUMMIT (1992) in 2012 described goals of sustainability by emphasizing both social and environmental concerns for sustainable development in any business.

(c) *The Business Discourse:* This aspect involves profitability, productivity, and financial performance by managing environmental and social assets composing its capital. Its inclusion in sustainability actually keeps a check on the depletion of natural resources and maintains consumption rate below natural reproduction, or at a rate below the development of substitutes, thus maintaining eco-effectiveness and sufficiency. In short, business sustainability means staying in business for the long duration [11]. By definition, it means meeting the needs of the firm's direct and indirect stakeholders without adversely affecting the needs of future stakeholders. Natural resources deterioration and the presence of social inequalities are placed at the heart of concerns. This helps in exerting control over society and producing large-scale innovations with governmental assistance in attending to eco-efficiency.

After conceptualizing sustainability as involving three components (environmental, social, and business), another edge is now included, which is good governance [12]. This fourth edge actually controls or oversees all the other three components. Therefore, sustainability requires all four dimensions to work effectively, through responsive and effective management. The most complex aspect is to balance all the dimensions of sustainability at an optimum level, due to a number of complications. For example, tensions exist between social and financial performance in which the former requires freedom and flexibility from financial constraints and business logic to find appropriate solutions for social problems, while the latter works in the opposite way to apply pressures on customers, employers, suppliers, beneficiaries, partners, and investors. Similarly, in the case of financial and environmental performance, the former works on a large scale to meet commercial needs with increased consumption of resources, which is detrimental to environmental performance. Therefore adequate management is required to normalize all tensions existing between different components of sustainability, through an integrated approach. This can be achieved by three effective measures:

- *Integrated Governance System:* This system requires compliance with national and international rules, regulations, and recommendations; integrated sustainability (social, environmental, and financial performance); effective risk management (holistic approach toward quantifiable and nonquantifiable risks, providing strategic perspectives, and decision making), and knowledge management (provide link to combine all the dimensions by development of skills, competencies, cultural and ethical support). The need for aligning a governmental system in sustainability is to ensure effective accountability for different stakeholders on national as well as international platforms. This combines achievement of financial and nonfinancial objectives, ethical behavior, risk awareness, and environmental concerns. Governmental principles and practices have helped in maintaining a formal structure for the process in the operational state [13]. Most importantly, management of risk and knowledge through regulatory power and legitimacy play a crucial role in sustainability management.
- *Hybrid Business Models:* These models have ensured the resolution of tensions created by various business models on environmental and social performance. These models have been adopted in various sectors due to the better optimization of various sustainability components. Based on this concept, even hybrid organizations have emerged among high-tech R&D firms involving joint efforts and collaboration between industry and academia. Further, the manager doesn't have to maintain a balance between social, environmental, and financial performance. Here, a virtuous cycle of long-term financial results and reinvestment is maintained through a delicate balance between social, environmental, and economic objectives. However, being hybrid in nature, this type of model can face some challenges when the mission drifts toward one direction and is detrimental for another direction. It can be a bit unstable in that it does not allow one-off solutions to be found, but rather requires a continuous

search for the "hybrid ideal." These models have the ability to address some of the critical challenges in integrated sustainability, which makes them desirable for managing all the components of sustainability [14].

- *Integrated management, measurement, and reporting systems:* These systems are important due to their ability to consider a broader spectrum of management controls involving several factors such as a firm's capital, governance structures, and business models, as well as performance drivers and outcomes. Further, they are actively involved in the value creation process and its interactions with different components. They are based on logistics rather than on sums or systemization of financial, social, and environmental reporting systems. They provide an opportunity for a broader rethinking of all preexisting systems based on either discrete or integrated sustainability or their compliance [15]. The integrated report should have an organized alignment with other measurement systems like business plans, balanced scorecards, budgeting systems, quality and production efficiency management systems, etc. Further it should be an active and constructive element, allowing an adequate planning process, enacting, monitoring and communicating at all organizational levels in integrated sustainability.

5.3 Implications of combo-technology for sustainability

The dominant drivers for bringing combo-technology into the life of ordinary people include keen business interests, improvements in economic status, and bringing solutions to the energy crisis as well as environmental remediation. Crucial factors such as governmental policies, business interests, consumers, involvement of nongovernmental organizations (NGOs) and self organizing groups (SOGs), as well as education and active awareness, are discussed in detail in the following paragraphs.

(i) *Government:* It needs to be involved at the highest level, being crucially important for bringing any technological changes, including implementation of combo-technology to solve various problems, as discussed in the previous chapters. Governance systems are operative at all societal levels (from global to local) it works in compliance with effective subordinateness. Technological success relies completely on transparency of governmental systems and policies, ones that are not prone to corruption and that maintain a balanced state with respect to short- and long-term interests for every class of society. Government plays a key role in regulating against adverse technological consequences by overseeing investments in research and development (R&D), purchasing sustainable products/services, setting parameters for fostering sustainable technologies, optimizing private interest-driven research with affable communications with the public at large, and finally setting long-term goals [16].

(ii) *SOGs and NGOs:* These organizations are a connecting link between the common people and the government. They interact directly with people, and thus are completely aware of their demands and problems. Technological success depends on public awareness, approachability, and ready adaptability. All three can be achieved once there are direct interactions with the people and it can be quickly seen whether or not the newly introduced technology is a success or a failure. However, there are various controversial issues associated with technologies introduced in the past, viz. nuclear energy and genetically modified foods in Europe and the United States. In fact, these organizations provide (R&D) inputs based on their surveys of new technology introductions, which can help in making further improvements in the technology. They also help in expressing common demands to governmental systems and people connected to business, accelerating and easing the introduction of the useful and most desirable technologies into the common marketplace. Sometimes these organizations are themselves involved in technological innovations using their own collected funds [17].

(iii) *Consumers:* They are the key factors in making any technology a real success. Successful introduction of any technology into the lives of ordinary people relies on making their lives better, which is the case with combo-technology. However, sustainability of technology also depends on the types of lifestyles of common people [18]. Sometimes, a technology does not prove to be sustainable, not because of technological issues but because of lifestyle issues. More consumer platforms have now been introduced, most importantly social media, to enable consumers to express their demands and preferences for products/services. This will be a factor in initiating appropriate innovations for upcoming technologies and thus will be an important determining facilitator in technological success.

(iv) *Business:* It can be divided into groups: big multinational corporations (MNCs), small and medium-sized enterprises (SMEs), and emerging new firms (generally science-based/service-oriented). The success of any introduced technology depends on the manner and timing by which it enters the market and fulfills sustainable needs of consumers, besides generating a modest profit for the business. It has been found that in the business grouping described, flexibility usually comes from emerging new firms, followed by SMEs and then MNCs. This is so because very large profits are not expected at the initial stage of any technology introduction into the market. A number of changes have been introduced at various levels so that technological innovations can be given enough time to sustain and prove their worth. The world governance system has regulated financial markets by applying standard taxation on all worldwide financial transactions. This will help companies to develop long-term sustainable strategies for their products/services/labor operations. Further, the World Trade Organization in association with the World Court and some parts of the United Nations have developed a socially motivated world governmental system that is firmly committed to sustainable development in addition to maintaining justice and equity. This global system has issued global standards of labor, environmental and social sustainability, and reasonable rather than excessive profits for all levels of business groups [19]. Further, intellectual property rights (IPRs) and the patent law systems (especially to lessen the amount of money spent in patent litigation) have been modified to promote more and more companies to adopt technological innovations as their business interests.

(v) *Education and Communication:* Complete information on any introduced technology should be provided, including its achievements and ambiguities. Awareness can best be started from high school and college education with emphasis on science, technology, and sustainability. Understanding the history of technology has been very helpful in the social shaping of technological artifacts, societal processes, and the decision structures shaping technological innovations. Further, the consequences of technology for society have to be emphasized for deeper understanding of technological change processes. Factors such as sustainability needs, correlation of technology with institutions and values, ecology with economy and society, consumers with producers and governments, short- and long-term goals in addition to well-being with equity, as well as differences between cultures with global values. The media/social media is another platform for contributing toward awareness on introduction of any new technologies in different sectors [20].

Technology development depends on effective interactions of factors such as scientific discoveries, business interests, consumer demand, government regulation, the global citizens movement, emerging institutions and paradigms, and ultimately changing dominant values. Technological innovation toward attainment of sustainability is the most challenging aspect to be achieved. It is the utmost responsibility of the scientific community developing any technology, such as combo-technology in the present context, to be completely aware of all the positive as well as negative aspects related to society, rather than being biased and working single-mindedly to place the technology into commercial markets. There should be an attitude of continuous development of better methods and motivating coordination between governments and academia, in addition to having firm determination regarding ethics and social responsibility. It has been suggested that more people (on local, regional, and global scales) should be involved in decision-making processes, creating the conditions for direct participation of more stakeholders in any released technology [21]. Sustainable development of the released new technology must meet the basic needs of all people and provide better opportunities to everyone, with achievement of a better life. Technology should not impeded the basic needs (food, clothing, shelter, and jobs), which is particularly important in developing and undeveloped countries. Briefly, sustainable development of any new technology is based on:

- Living standards should be heightened and longer lasting.
- Consumption standards should be within the limits of ecological possibility.
- Increased production potential should be emphasized along with ensuring equal opportunities for all.

- Demographic developments should be in harmony with the changing productive potential of the ecosystem.
- Human activities should not endanger natural systems (support life on Earth, atmosphere, waters, soils, and other living beings).
- Thorough knowledge of the released technology, including both positive and negative aspects, should be made available.
- The technology should attract people from various business sectors.
- There should be no overutilization of natural resources to meet ever-increasing demands.

Combo-technology has provided solutions for various environmental problems, and thus is a powerful tool for sustainability. The best part of this technology is neutralizing the toxicological effects raised by nanotechnology on the whole ecosystem when released into the environment. Nanotechnology undoubtedly brings some solutions for environmental problems like the energy crisis and environmental pollution. However, nanomaterials with their very small size can be easily mobilized inside living cells and induce toxicological effects. They have very high cost, especially when used in pilot and field trials. There are incomplete ecotoxicological profiles for various nanomaterials with little known about their potential health and environmental concerns [22, 23]. It has been estimated that an average of 15 years would be required for proper analysis and validation of nanomaterial risk assessments. There is no regulatory strategy at the present time to address risk issues of nanotechnology and their defined protocols for testing before their release into commercial markets. The most important concerns raised about nanotechnology are the huge energy and environmental costs during manufacturing of nanomaterials, and toxicological effects of nanomaterials on the environment and human health. Intensive research on factors responsible for various concerns raised during usage of nanomaterials have revealed that size, shape, surface charge, composition, coatings, medium, and surface roughness are the real determinants of nanomaterial toxicity [24, 25]. Enzyme technology has provided some solutions in this respect, in which nanomaterials are used as matrices for immobilizing enzymes on their surface. Thus, factors like size, shape, surface charge, and composition are nullified, and beneficial enzyme properties are also added. Enzymes are available from a wide range of sources (prokaryotes to eukaryotes), with extraordinary specificity, having several thousand times higher catalytic turnover. They can work under milder conditions using traditional chemicals (nontoxic). They are easily biodegradable with no toxicity to environment in any respect [26]. Environmental assessments of enzymes by LCA (lifecycle assessment) and EIA (environmental impact assessment) have strongly recommended their usage for wide applications. Enzymes have taken over a global market of about $1000 million in 2016 and are expected to increase cumulatively in the coming years. Economists have firmly declared that there is a strong correlation of urbanization and GDP increase with the growing enzyme market. Previous chapters have thoroughly discussed the fact that immobilized enzymes based on nanomaterials have been useful in producing clean energy and remediating contaminated air and water. Immobilized enzymes based on nanomaterials can produce energy using very cheap raw materials like lactose, sucrose, cellulose, xylan, steam exploded aspen wood, starch, etc. in a cost-effective way without generating any greenhouse gases. Immobilized enzymes can effectively transform highly toxic, recalcitrant pollutants (polycyclic aromatic hydrocarbons, petroleum hydrocarbons, phenols, polychlorinated biphenyls, azo dyes, organophosphorus pesticides, heavy metals, etc.) in air, water, or soil into less toxic or innocuous products. Further, nanomaterial-based immobilized enzymes are also helpful in checking the level of contaminants in air and water, as they are important components of air and water sensors. Enzyme immobilization in general helps in lowering the production cost by several folds due to enzyme ability to be reused several times without much loss in catalytic efficiency [27]. It is the simplest technique, with convenient handling of enzyme preparations. Immobilization yields easier enzyme separation from the product, making it compatible for a wide range of applications, including those discussed in previous chapters, and thus minimizing downstream product processing. In the case of nanomaterial-based immobilized enzymes, the enzymes are also involved in making interlinked cross-bridges between the nanomaterials, besides being present on their surfaces. This helps in prevention of any leakage of nanomaterial and even the enzymes themselves to contaminate the final treated product, i.e., air and water, as well as in production of bioethanol, biodiesel, hydrogen, etc. Therefore, consumers are never in direct contact with either nanomaterials, which have several issues related to environmental and health safety, as previously discussed, or enzymes, making the immobilized enzymes (based on nanomaterials) completely safe for consumers. Combo-technology has completed fitted all the parameters together as set by sustainability concepts, including cost efficacy, environmental friendliness with great social impact without wasting any natural resources, involving people from various sectors with varied expertise (providing a platform for a wide range of employment opportunities), and providing solutions to most severe environmental problems. The beneficial features of immobilized enzymes have created enthusiasm among industrialists to maximize the use of immobilized enzyme–based processes. It will be even more interesting when many novel enzymes are discovered with specificity for a number of recalcitrant compounds present in the environment, thus giving complete solutions for many environmental problems.

5.4 Analyzing environmental sustainability index of combo-technology

5.4.1 The concept of environmental sustainability index

The Environmental Sustainability Index (ESI) is the measure of environmental sustainability among various countries around the world. The ESI is the most powerful and useful measure of conditions, specifically focusing on current societal performance and capacity for future policy interventions that determine long-term environmental trends. It is based on 20 indicators in addition to a combination of two to eight variables, which results in a total of 68 data sets. No country in the world is fulfilling all 20 indicators, as there are some issues leading to incompetency in achievement of complete sustainability [28]. It has been correlated that the higher the ESI score, the better are the environmental conditions prevailing in the respective country. The five countries found to have the highest ESI scores are Finland, Norway, Sweden, Canada, and Switzerland, while the five countries with the lowest ESI scores are Haiti, Iraq, North Korea, Kuwait, and the United Arab Emirates [29]. The complete 20 indicators that comprise the ESI have been correlated with per-capita income with considerable variations. Rich countries have higher ESI scores based on social and institutional capacity, current ambient conditions, as well as on reducing human vulnerability, with exclusion of factors like land and biodiversity. However, less wealthy countries can also generate better ESI scores due to lower environmental stress with good amounts of production of biodegradable wastes (due to high population density) with exclusion of gaseous emissions [30]. Therefore, a country's wealth cannot be given (related to income) much weight toward achievement of good ESI scores. This can be explained as follows: a country like Korea has a higher score than Portugal (due to higher levels of science and technology); Sweden has a higher ESI score than Italy, while Estonia has a higher score than Saudi Arabia, despite the fact that each pair of countries has similar levels of GDP per capita. Thus, each indicator of ESI is associated with a number of different variables being empirically measured. The complete database has been constructed based on the extent of particular data coverage, total population of the country, and the size of its territory. The countries with population $<100,000$ having $<5000\,km^2$ of area, and lacking sufficient data to generate indicator values, have been eliminated and a total of 142 countries have been included, to generate 68 factors/variables corresponding to 20 indicators in calculating ESI scores with compensation of about 9656 missing data points (i.e., 22%) in the database. Regarding ESI scores of the 142 countries, a bell-shaped curve has been generated with Finland positioned at top right, Nigeria at bottom left, with Belgium at bottom right and Uruguay at upper left, etc. [7]. Strengthening ESI factors are: effective governance, good economic and environmental management, effectual economic strategies, negligible corruption, civil liberties, and effective democratic institutions. Further, the indicator "private sector responsiveness" has been also corroborated, which helps in developing innovative responses toward environmental challenges. However, the indicator "climate change" has a negative correlation with economic competitiveness and environment sustainability, with factors like high pollution levels and rising greenhouse gas emissions leading to worsening effects.

ESI scores help in assisting countries toward making proper environmental decisions, particularly by administrative authorities. The five core components of ESI are: improvement of environmental systems, reducing stresses and human vulnerability, social and institutional capacity, and global stewardship. The ESI enables identification of issues related to national performance (above or below expectations), tracking of environmental trends, priority-setting among policy areas within countries and regions, quantitative assessment of policies and programs, probing interactions between environmental and economic performance, and responsible factors for environmental sustainability. ESI is rightly a linking factor between per-capita income and level of development [31]. A comprehensive index based on complete profiles of pollution control and natural resource management issues would reveal the impoverished state of environmental metrics across a large number of countries of the world. The ESI is an effective metric for gauging prospects for long-term environmental sustainability based on current conditions, pressures on those conditions, human impacts, and social responses as well as underlying resource endowments, past practices, current environmental results, and the capacity to cope with future challenges. The ESI has permitted more accurate analysis of the correlation between economic competitiveness and environmental sustainability. The World Economic Forum 2001 has also confirmed this by showing a correlation of 0.34 (statistically significant) between current competitiveness index and ESI-based data obtained from 71 countries [32]. Various important variables have been identified to have significant correlation with the ESI: governance (including three independent data sets of civil liberties, democratic institutions, and control of corruption), geography (negative correlation between ESI scores and population density), and other geographical factors (including distance from equator and climatic zones) [33]. ESI and EPI (Environmental Performance Index) have been useful in accurate assessment of various schemes and technologies involved in pollution control and natural resource management (an overview on EPI has been given in Fig. 5.1). The development of ESI requires thorough analyses involving an open and interactive process, statistical analyses, as well as environmental and analytical expertise to analyze data obtained from different parts of the world. It is difficult to measure things in the environmental domain (mostly based on empirical underpinnings without sufficient analytical rigor) but it is of the most urgent necessity if correlation is to be made with business, in which every aspect is measurable. The ESI has been very useful in making the concept of environmental sustainability more concrete and functional using reliable metrics. It has set baseline conditions and priorities for establishing suitable targets, identifying trends, and understanding the actual determinants of policy success. Environmental issues like climate change, deforestation, and ozone depletion have been carefully addressed based on ESI scores. However, some environmental sustainability issues are still not being dealt with due to lack of complete data. The basic unit of environmental sustainability has been presented as a function of five phenomena: the state of the environmental systems (such as air, soil, ecosystems, and water), the stresses on those systems in the form of pollution and exploitation levels, the human vulnerability due to environmental change (viz. loss of food resources, exposure to environmental diseases, etc.), the social and institutional capacity in grappling with environmental challenges, and the ability to respond to the demands of global stewardship through cooperation [34].

The set of 20 indicators in the ESI have resulted in improvements in most of the environmental conditions and more economic development, as well as simplifying to a large extent the complex relationships between economic and environmental outcomes. Further, the welfare effects of trade and investment liberalization have been limited by the paucity of environmental data held up against the abundant economic data. The ESI has significantly contributed toward measurable parameters of environmental sustainability, narrowing the major gap present in the arena of environmental policy, executable environmental goals and measuring their progress and performance,

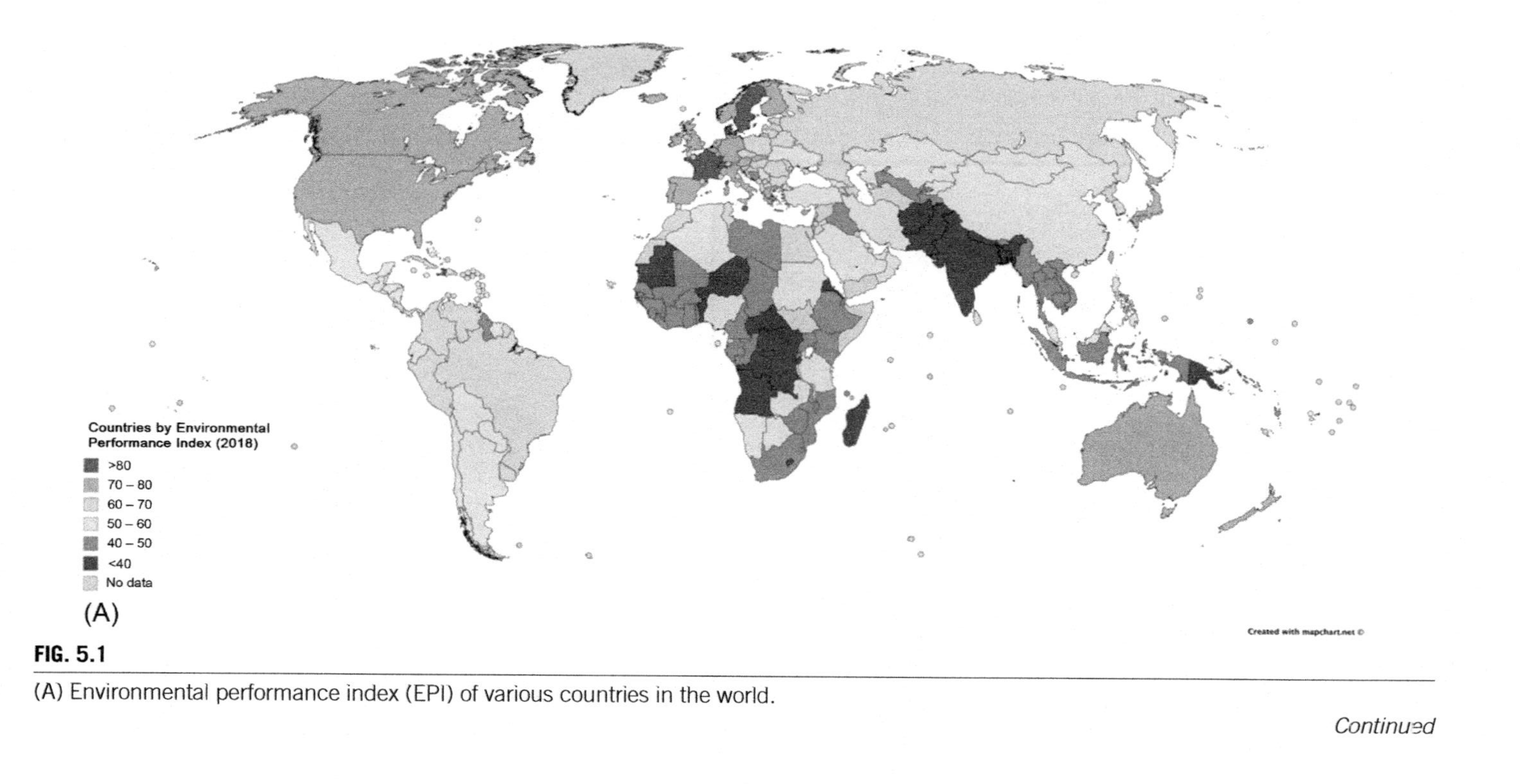

FIG. 5.1

(A) Environmental performance index (EPI) of various countries in the world.

Continued

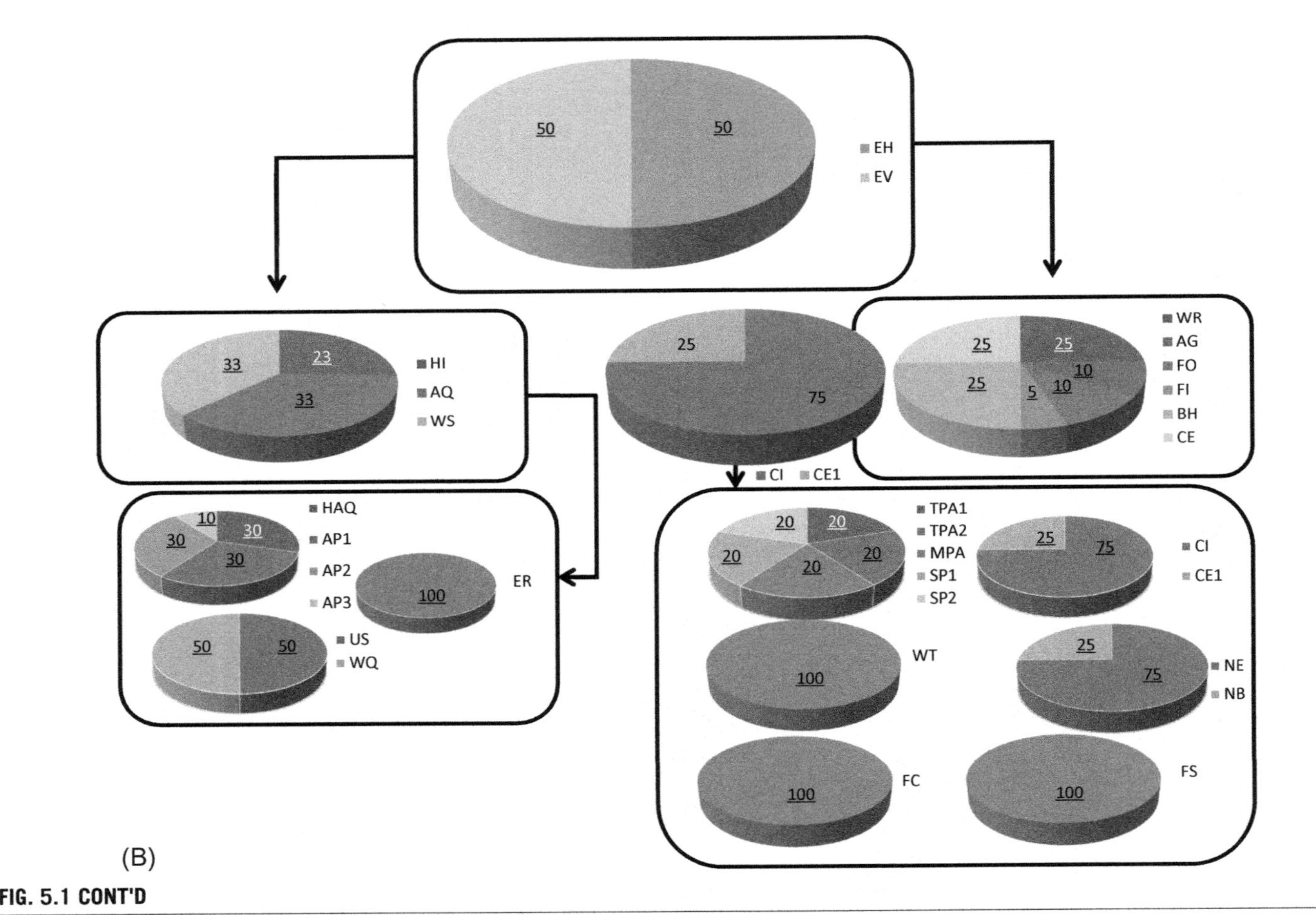

FIG. 5.1 CONT'D

(B) Environmental performance index (EPI) is based on various indicators as schematically represented along with their share percent. *AG*, agriculture; *AP1*, air pollution: average exposure to $PM_{2.5}$; *AP2*, air pollution: $PM_{2.5}$ exceedance; *AP3*, air pollution: average exposure to NO_2; *AQ*, air quality; *BH*, biodiversity and habitat; *CE*, climate and energy; *CE1*, trend in CO_2 emissions per kWh; *CI*, trend in carbon intensity; *EH*, environmental health; *ER*, environmental risk exposure; *EV*, ecosystem vitality; *FC*, change forest cover; *FI*, fisheries; *FO*, forests; *FS*, fish stocks; *HAQ*, household air quality; *HI*, health impacts; *MPA*, marine protected areas; *NB*, nitrogen balance; *NE*, nitrogen use efficiency; *SP1*, species protection; national; *SP2*, species protection; global; *TPA1*, terrestrial protected areas; national biome weights; *TPA2*, terrestrial protected areas; global biome weights; *US*, unsafe sanitation; *WQ*, water quality; *WR*, water resources; *WS*, water and sanitation; *WT*, wastewater treatment.

(A) From https://en.wikipedia.org/wiki/Environmental_Performance_Index.

facilitating refined investigation of the drivers of environmental sustainability, and focusing attention on best practices in environmental decisions. It has offered both aggregate ranking and disaggregated data to accurately calculate environmental analysis at various scales, providing a database helpful in building transparent, reproducible, and refined methodologies, and maintaining the balance between the need for broad country coverage and the need to rely on high-quality data that are often of more limited country coverage. However, there are a few shortcomings related to ESI scores: they rely in some instances on data sources having limited country coverage and poor quality; they lack time series data; they have substantive gaps in a number of high-priority issues; some indicators implying high priorities and values have not been universally shared; and extensive validation is missing for identifying empirical determinants of good environmental performance [35]. Experts are still working on ESI to come up with an improved version by undergoing several analytical refinements to deepen understanding of environmental sustainability and its measurable parameters. There is a need for a pluralistic approach in filling critical data gaps by using existing international organizations, networks of scientists, local and regional officials, industries, and nongovernmental organizations. There should be flexible information systems permitting users to apply their own value judgements or to experiment with alternative casual hypotheses. ESI should be made very interactive with easier operability on any computer with complete transparency of available data. More and more powerful tools should be provided for effective ESI data integration from different sources. The phenomenon of environmental sustainability is a function of the interaction of mechanisms that operate at various levels, like ecosystems, watersheds, firms, households, and economic sectors. Thus, there should be continuous efforts for information integration at different spatial scales, evaluation, and supplementation. Further, robust investigations should be carried out at different time intervals for studying cause-effect relationships as well as for improvement of data aggregation. The process of data evaluation should be continued into the future, with the possibility of retrospective measurements of certain variables which are given causal analysis.

5.4.2 Nanotechnology firms vindication toward high ESI

Nanotechnology has wide applications in every sector from its utilization in design, characterization, production, and application of its materials, structures, devices, and systems. It is not just a future technology but has already found a place in our present and improved our lives through its various tools being used, both at home and in the workplace. Nanotechnology has become an influential factor in the global economy, expected to reach \$90.5 billion by 2021 from \$39.2 billion in 2016 with CAGR (compound annual growth rate) of 18.2%. A CAGR of 28.2% has been reported in the case of the nanodevice market, with expected increment from \$56.5 million in 2016 to \$195.9 million by 2021. The market share of nanomaterials is expected to increase to 85.3% by 2021; however there will be a decrease in the case of nanotools to 14.5% with slight increment of 0.1% in the case of nanodevices. It has been reported that nanotechnology occupied its largest market share in 2015, with 38.8% (of total market) in environmental applications, 22.4% in electronics, and 21.1% in consumer applications [36, 37]. In the same year, CAGR rates as reported corresponding to biomedical, consumer, and electronics applications were 29.9%, 27.9%, and 20.5%, respectively. Nanotechnology has been strongly promoted by almost all countries of the world. The European Commission has invested 896 million euros in nanotechnology-related research between 2007 and 2011. Presently, nanotechnology is worth 1 trillion USD (quarter of worldwide investment) and is still growing.

The BRICS (Brazil, Russian Federation, India, China, and South Africa) countries have become a hub of nanotechnology. Several articles related to nanotechnology have been published in addition to several patents registered by BRICS countries. The ratio of patents to articles published in nanotechnology has been calculated in 2015 as: 2.47 per 100 articles for South Africa, 2.28 for China, 1.67 for Brazil, 1.61 for India, and 0.72 for the Russian Federation. However, in case of Italy, United Kingdom, and Canada, it has been found to be 4.46, 8.39, and 10.08, respectively. Brazil represents 1.6% of total world output in nanoscience and has been making major investments through federal and state funding in nanoscience research projects. China has been long known for its traditional strengths in materials science, chemistry, and physics. Thus, it has been very successful in establishing nanotechnology quickly and has been able to come up with various nanotechnology-based products in the market. China has the greatest research intensity in nanotechnology with respect to other BRICS countries, having significant contributions in enhancement of GDP: 2.08% (China), 1.15% (Brazil), 1.12% (Russian Federation), 0.82% (India), and 0.73% (South Africa) [38]. In India, the development of nanotechnology has been oriented more toward building human capacity and physical infrastructure rather than in products commercialization. India has only launched two products based on nanotechnology, related to personal care, as reported by a consumer products inventory. However, India has been working to become a global knowledge hub in nanotechnology. It has invested funds in 240 research projects related to nanotechnology and has established several institutes dedicated to nanotechnology research. Further, the government has set up various nano-manufacturing technology centers within the existing Central Manufacturing Technology Institute to strengthen public-private partnership. In the Russian Federation, there are over 500 companies manufacturing nanotech products with the sales of more than US$ 15 billion (as reported in 2013). Almost one-quarter of Russian nanotech products, worth US$ 2.30 billion, are exported. The Russian Federation has started Rusnano (projects related to nanotechnology), which has supported over 98 projects and established 11 nanocenters for technological development and transfer; and four engineering companies in different regions. Nanotechnology has become the country's sixth research priority for civil-purpose applications, after transport systems and space (40% of total funding), safe and efficient energy systems (16%), information technology (12%), environmental management (7%), and life sciences (6%). Lots of nanomaterials have been developed with various applications; however, there are very few academic articles or patented inventions related to nanotechnology. Malaysia and Iran have come up with good growth in the number of articles being published in nanotechnology. Iran stands 7th and Malaysia at the

22nd rank in the number of articles published related to nanotechnology. The ratio of nanotechnology patents to total 100 articles published was found to be 0.41 and 0.73 for Iran and Malaysia, respectively, in 2015. Iran set up the Nanotechnology Initiative Council (NIC) in 2002 for the development of R&D in nanotechnology and provides facilities for creating markets for the private sector. In the past decade, 143 nanotech companies have been established with more than one-quarter in health care and just 3% in the automotive industry. The Nanotechnology Research Centre has been established at Sharif University, which has started Iran's first doctoral program in nanoscience and nanotechnology. Further, Iran has hosted the International Centre on Nanotechnology for Water Purification in collaboration with UNIDO in 2012. In 2008, an Econano network was set up to promote scientific and industrial development of nanotechnology among Afghanistan, Azerbaijan, Kazakhstan, Kyrgyzstan, Pakistan, Tajikistan, Turkey, Turkmenistan, and Uzbekistan (all are fellow members of the Economic Cooperation Organization). There is a growing interest in nanotechnology in other countries like Argentina, Chile, Croatia, Jordan, Mexico, Morocco, Nepal, the Philippines, Saudi Arabia, Serbia, Slovenia, Sri Lanka, and Tunisia. Of these countries, only Slovenia has devoted >1% of GDP to R&D in nanotechnology. Nepal has planned to set up the National Nanotechnology Centre in the coming 5 years. In the Philippines, the Centre for Nanotechnology Application in Agriculture, Forestry and Industry has been established at the University of the Philippines, Los Baños in 2014. This center is active in providing structural framework and support systems for the ownership, management, use, and commercialization of intellectual property through government-funded R&D. The InnovAct program was started in 2011 by Morocco, which has provided up to 30 enterprises per year related to nanotechnology and other strategic fields with logistical support and the financial means for various research projects. King Abdulaziz City for Science and Technology (KACST) in Saudi Arabia has been set up for improving ties between public and private sectors, particularly in the area of nanotechnology and advanced materials. SLINTEC (Sri Lanka Institute of Nanotechnology) in Sri Lanka was established in 2008 (a joint venture between the National Science Foundation and Sri Lankan corporate giants like Brandix, Dialog, Hayleys, and Loadstar). It has aimed to set up a platform for commercializing nanotechnology-based products with specialization in smart agriculture (viz. nanotechnology-based slow release of fertilizers), rubber nanocomposites (viz. high-performance tires), apparel and textiles (viz. smart yarns), consumer products (viz. detergents, cosmetics), and nanomaterials. The Nanotechnology and Science Park and Nanotechnology Centre of Excellence were established in 2013 for providing high-quality infrastructure for nanotechnology research in Sri Lanka. The ranking of countries like Iran, India, Pakistan, Sri Lanka, Nepal, and Bangladesh as per number of articles published in nanotechnology per million populations are: 27th, 65th, 74th, 83rd, 85th, and 90th. Arab countries are also looking forward toward development in nanotechnology through various plans. Governments have planned specifically to develop nanotechnology with applications ranging from health and pharmaceuticals to foodstuffs, environmental management, desalination, and energy production. There is a firm emphasis in developing human resource and international cooperation through nanotechnology-based developments. UNESCO has promoted the development of linkages between academia and industry as well as removal of barriers hindering innovation in the Arab world. NECTAR (Network for the Expansion of Convergent Technologies in the Arab Region) has been set up in collaboration with renowned Arab scientists based at universities in the United States and Egypt to modernize university curricula and convergent technologies, including nanotechnology in the Arab region. NECTAR has also started various diploma courses in addition to providing a masters degree in nanosciences. Graduates coming from NECTAR are being recruited in industries based on pharmaceuticals, chemicals, petrochemicals, oil production, optoelectronics, electronics, information technology, fertilizers, surface coating, building technology, foodstuffs, automotive industry, etc.

The most important features of nanotechnology responsible for its good ESI are:

(i) *Economic crisis and recovery through innovation:* The European Union (EU) has recovered from the major global economic crisis in 2008–10 originating from economical imbalances accumulated in the period 2000–07, mainly due to inflation of house and stock prices. Most of the EU member states recovered by mid-2009, while member states like Greece, Ireland, and Romania recovered later, i.e., by the beginning of 2011. The level of recession ranged from a tiny one-quarter drop in Poland to 25% loss in Latvia, Ireland, and Spain. Countries like Austria, Belgium, and Germany have suffered recession due to collateral effects from readjustments in the United States and the first group of EU member states, chiefly affected through international trade and their financial systems to loans. The solution has been found to focus intensely on R&D and innovation, as it is the most important source of sustained growth [39] (Fig. 5.2). Several strategies have been built up to ameliorate R&D, viz. the Europe 2020 strategy to maintain the Lisbon target of 3% for R&D intensity for the EU as a whole, rather than taking individual member states. It has been important to develop R&D innovation in the United States due to collateral relations. Major commercialization of R&D has been an intensely focus. However, selecting the key areas that can produce faster results with successful applications is not always easy. Several sectors of basic research have been fostered to create new ideas and foster favorable business conditions for new technology and innovation, which can be commercialized into the market. Further, this complete process should be well complemented by an adequate level of intellectual property rights protection, to provide enough incentives to the innovators to promote more new ideas. Nanoscience and nanotechnology have become the research priority for most of these countries, due to its wide applications in every sector of our daily lives [40]. Switzerland has excelled in having the highest output in nanotechnology, thus has topped the EU's Innovation Scoreboard and Global Innovation (2014) as well as being one of the top three countries for innovation among members of OECD (Organization for Economic Co-operation and Development). Switzerland published 198 scientific articles per million population in 2013 and scored the highest record with respect

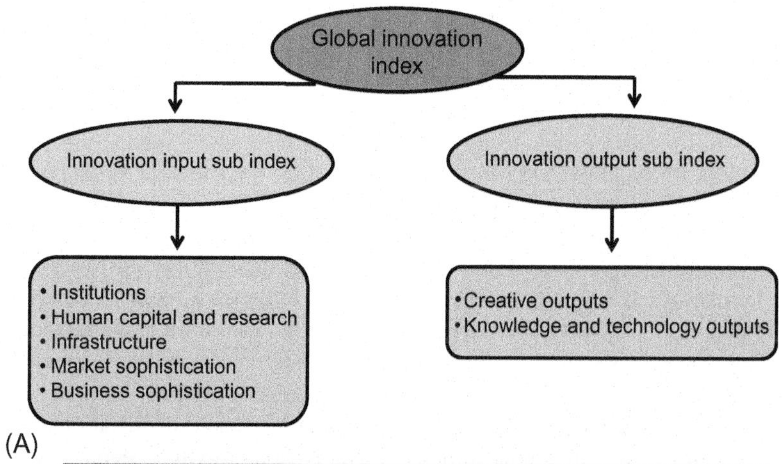

(A)

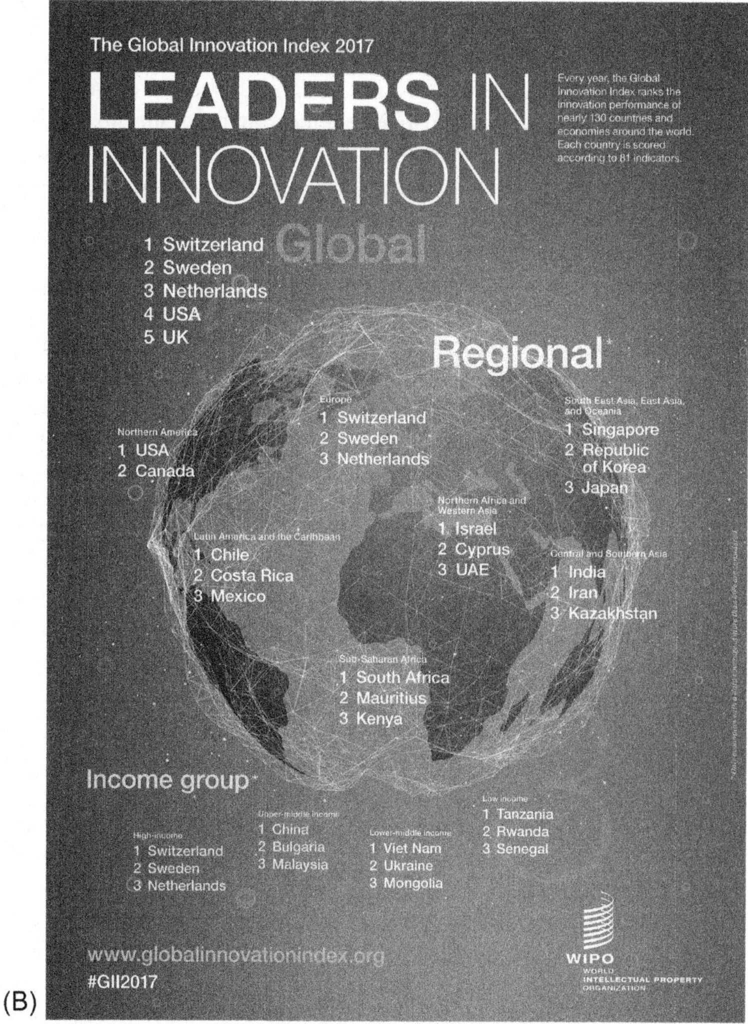

(B)

FIG. 5.2

Advancements in the technology can be calculated based on GII (Global Innovation Index). (A) Schematic representation of GII composition. (B) Leaders in GII around the world based on their economic status and location.

Global Innovation Index 2017. Innovation feeding the world (Cornell University, INSEAD and WIPO). Infographic reproduced with permission of WIPO.

to other strong players at lower positions like the Republic of Korea (150), Germany (93), France (79), the United States (69), and Japan (56). However, with respect to the number of patents per 100 articles on nanotechnology, the positions are United States (44), Japan (30), the Republic of Korea (27), Germany (22), Switzerland (17), and France (15). The Wroclaw Centre of Excellence is known to be the first team project to work on nanomaterials, particularly nanophotonics involving collaboration between the German Fraunhofer Institute for Material and Beam Technology, University of Würzburg, Wroclaw University of Technology, and the Polish National Centre for Research and Development. Nanotechnology has become a key element in advanced manufacturing and is being pursued by a larger number of countries on a grand scale, including Australia, Canada, China, France, Germany, Japan, the Republic of Korea, and the United States [41]. The government of China has started 16M-engineering programs up to 2020 for advanced manufacturing. Canada has revised its research strategy with a focus on automation (robotics, lightweight materials, and technologies), quantum materials, and nanotechnology. Further, the Canadian government is hoping to enhance national competitiveness and jobs avenues through advanced manufacturing. An advanced manufacturing partnership was launched by the United States in 2013 involving industrial, labor, and academic sectors with an investment of US$2.9 billion under the Revitalize American Manufacturing Act (2014). The funding has been used to start up 15 institutes focusing on additive manufacturing, like three-dimensional printing, digital manufacturing and design, lightweight manufacturing, wide-band semiconductors, flexible hybrid electronics, integrated photonics, clean energy, and revolutionary fibers and textiles. Japan has taken the sixth rank worldwide in publishing nanotechnology-related articles with huge industrial investment in nanotechnology of ¥111 billion in 2013. In the case of university funding for nanotechnology, it was about ¥55 billion in 2013 (below industrial levels). However, Japan is one of the rarest cases showing a decline in the volume of scientific articles published over the past decade, which has led to a shrinking of Japan's share of articles in nanotechnology on the world platform.

(ii) *Drive Prosperity and Global Competitiveness:* It has been difficult to predict the impact of emerging technologies on business and economics on the basis of their present developments. That is true even for nanotechnology, despite several claims and achievements. However, nanotechnology in its present state has been helpful in reshaping the global economy, with a potential of realigning society, changing business, and affecting economics at a structural level. Various business models and manufacturing strategies have evolved having excellent efficiency and reduced price points [42]. Nanotechnology has impacted various economical aspects such as wages, employment, purchasing, pricing, capital, exchange rates, currencies, markets, supply and demand. Nanotechnology has been quite effective in evoking economic prosperity, shaping productivity, and increasing global competitiveness. It has brought several marvelous innovations, such as reducing production cost of computers by 50%, reducing manufacturing cost in drug development by 70%, and even has been known to be the best substitute for fossil fuels by providing clean energy, etc. It has reduced the cost of various essential goods and services in areas like quality of life, habitat, and transportation. However, despite these fruitful results, it is still very difficult to predict the future related to nanotechnology due to the complex interlinkage of markets, industries, and economy. It appears that innovations in the coming decades will be built on nanotechnology and its derivates, based on predictions from the fundamental economic landscape of today's strong economy.

(iii) *Evolution of Nanoeconomy (basis of nations' future wealth):* Technical innovations have had great impacts in shaping countries' economies (driving global as well as domestic GDP) and market robustness. Nanotechnology with its added innovations through advances in its R&D appears to have the power to move nations' economies upward [43]. The nano-economy (nanotechnology-derived economy) will possibly bring about complete replacement of the petro-economy (fossil-derived economy). Economic transformation through nanotechnology is based on integration into industries such as health care, manufacturing, and energy. There are several challenges, even though things are moving very fast from theoretical to practical approaches and strong impacts are already being made on business, society, and the economy. Nanotechnology is a multidisciplinary field that requires collaborative research efforts and innovation networks at national and international levels. It would provide a platform for offering new approaches for education, innovation learning, and governance. Huge public investments have been made to support scientific and technological researchers in creating technological and industrial platforms as well as infrastructures, as shown in the over 2×10^6 articles related to nanotechnology that have been published, with over 10^6 applications filed with patent offices. The commercialized nanotechnology innovations have brought solid economic value for nations through funded research via supportive investment and workplace manufacturing environments [44]. This technology has taken <10 years to be transferred from West to East, due to its bulk manufacturing capabilities.

(iv) *Nanotechnology Innovation and Commercialization:* The usual trend being followed for commercialization of nanotechnology innovation involves an interlinking of large and small companies, research organizations, equipment suppliers, intermediaries, finance and insurance, end users (both in private and public sectors), regulators, and other stakeholder groups (usually present in a highly distributed global economy). Over 17,600 companies worldwide have published about 52,600 scientific articles and filed over 45,052 patents in the nanotechnology domain (in the period 1990–2008). Most of the countries of the world are actively involved in

> the development of nanotechnology, with the United States at the forefront of nanotechnology innovation. This is due to direct correlation between economic development and productive commercialized nanotechnology innovation. Novelty in technological innovation provides the safest ground for commercialization. The degree of innovation is based on either the creation of completely new products/services or just the improvement of existing products/services. Further, it should provide well-equipped experts to train individuals for elaboration of commercial platforms of the newly introduced innovations. The innovative process is not grounded on a single individual/firm developing it, but instead requires a network of individuals/firms for its complete success. The innovations in nanotechnology are increasing day by day, with contributions from not only wealthy and large countries but from small countries or even regions. In the present scenario, there have been many evolutions in regional science and technology as well as industrial/enterprise policies, to converge toward providing better support systems for technology success. Geographically, various parts of the world have been clustered (related to technology innovation) based on the territorial extent of the firms, customers, suppliers, support services, and institutions. These activity-rich clusters are involved in critical activities related to industries and firms in these clusters carrying out the core strategy-setting, product or service development, marketing strategy, and corporate coordination activities. However, activity-poor clusters just involve one or only a few activities in a given industry or set of related industries. Further, in the case of high innovation and low innovation clusters, they are based on levels of innovations with the former being more innovative and having an ability to sustain itself due to the excellent innovative capabilities.

The Institute of Global Futures has done a thorough assessment of the economic and business impact of nanotechnology through a series of interviews with a broad range of business executives involved in health care, manufacturing, medicine, real estate, information technology, consumer goods, entertainment, and financial services. The Institute for Global Futures has released a 10-year assessment on the impact of nanotechnology on markets, society, customers, and the economy by covering broad sectors of R&D, including telecommunications, robotics, computers, life sciences, the Internet, software, and artificial intelligence [45]. It has clearly indicated cumulative wealth generation (rate of growth of scientific production is ~10% per year) through a wide range of markets, research inclinations, and supportive administrative frameworks. Systematic investment in research on nanotechnology has been undertaken by the United States since 2001, the EU since 2003, Japan since 2006, and other countries since at least 2005. Nanotechnology has set an example for sustainable development through increased productivity, improved health, and friendly association with nature. Nanotechnology has proven a platform technology with a potential to transform wide ranging industrial sectors through nurturing the convergence between previously separated technology-driven industries. The interdisciplinary nature of nanotechnology has included various scientific developments across various disciplines, which provide newness and asymmetric dispersion of knowledge. Further, a strengthening network of individuals, firms, universities, research institutes, venture capitalists, and public policy agencies has developed. Nanotechnology has been successful in translating into sustainability of nations, organizations, and entire industries with its nimbleness in preparation, planning process, and preparedness [46]. Thus, it has been effective in strategic planning as well as gaining widespread investments from business, education, labor, and government. Nanotechnology governance must be more open and transformative with an orientation toward funded projects based on technology innovations as well as looking intensely into EHS (environmental health and safety), equitable access and benefits, long-term planning, and anticipatory adaptive measurements. Further, governance of nanotechnology has focused on the first generation of nanotechnology products and their preparedness toward education, capital, talent, coordination, and communications. It should have been promoting national policy and investments leading to economic agility. Nanotechnology has been successful in giving a common platform for society and industry leading to improved national productivity and quality of life [47]. Collectively, nanotechnology has been helpful in global trade leadership, sustainable economic growth, higher productivity, robust gross national product, global patent leadership, superior industrial competitiveness, integrated education and training resources, plentiful capital liquidity, strong investment in climate, high investment in R&D, low unemployment, and excellent coordination between government and industry. Therefore, it clearly is justified in meeting all the parameters of sustainability, in turn leading to the achievement of excellent ESI scores.

5.4.3 Finishing touch to nanotechnology: a step toward making it sustainable over the long term, with sustainable excellent ESI scores

Actually, nanotechnology is a continuation of the next chapter in the acceleration of advanced technology providing a major platform in the transforming of the future of the global economy. It is the fastest emerging technology responsible for shaping of sustainability, economic progress, and is a basis of many industries in the present and coming years. With increasing applications of nanoparticles, there are serious issues related to environment and health. Nanotechnology has undoubtedly enormous applications in various fields including medicine and the environment; however, it is an important issue to consider the impact of nanomaterials on living organisms as well as on the environment. There is an easy entry of nanomaterials inside living organisms due to their small size and ability to evade natural trapping mechanisms of the body. Further, aggregation of nanomaterials inside the body has raised concerns due to severe complications, as discussed in Chapter 1. Various aspects of nanomaterials have been considered for their toxicity like size, shape, chemical composition, dispersion properties under various conditions, surface coatings, surface charge density, etc. Generally, nanomaterials ($\leq$10 nm) in the air are found to have the highest toxicological effects due to easier entry into the body [48]. Besides chemical composition, surface density has been found to be a crucial factor in determining toxicity as well as aggregating behavior. Therefore, size and surface density are being thoroughly studied to neutralize toxic effects of nanomaterials. As discussed in Chapter 1, surface charge density can be best utilized in enzyme immobilization, which adds a catalytic property to nanomaterials besides neutralizing toxicological effects. Most importantly, this has been effective in controlling aggregation behavior of nanomaterials. Besides neutralizing toxicological effects of nanomaterials, enzymes themselves gain various excellent characteristics like improved stability (by resisting protein unfolding), reusability (also helpful for cost reduction of the complete product), and enhanced activity. Nanomaterials are best suited for immobilizing enzymes due to their large surface area and small size that leads also to effective enzyme loading with reduced mass transfer resistance. It has been shown that enzymes immobilized onto nanomaterials show Brownian movement when dispersed in aqueous solutions, which imparts better enzymatic activities than soluble enzymes [49]. Enzyme-bound nanomaterials are much more stable and can be easily separated from reaction mixtures after their usage, thus making the complete process highly economical.

Government policymakers have taken the utmost interest in developing safe nanomaterials besides considering their wide applications. Existing governmental regulatory frameworks (30–40 years old) have been modified by focusing on toxicological effects of nanotechnology that are known or yet to be discovered. NIH has been working to provide a comprehensive database (complete information) of all the released nanomaterials, so that they can be managed accordingly. The negative aspects have been largely ignored in the previous decades when nanotechnology evolved, due to their wide applications and excellent market value. NIH plans to fund the research on interaction of nanomaterials with biological systems and biomolecules (proteins, DNA, and RNA) as well as environmental agents in its native states as well as modified states (viz. by enzyme immobilization). Further, NIH has also been developing some mathematical and analytical tools that can predict toxicity of released nanomaterials in both native and modified states [50]. This is a collaborative effort from mathematicians, chemists, physicists, biologists, and pharmacists. The Environmental Protection Agency (EPA) has found that, by 2020, the total global demand for nanoscale materials, tools, and devices would be ~$100 billion. Therefore, it is an utmost priority to perform the highest quality research in the areas of health and environmental effects of manufactured nanomaterials. The EPA has sanctioned several million dollars in funds toward thorough investigations of released nanomaterials, especially their bioavailability and bioaccumulation. The EPA has also planned to fund research on exploring more enzymes that are compatible with various synthesized nanomaterials and that produce negligible side effects on health and environment [51]. NIOSH (National Institute for Occupational Safety and Health) has been working on identifying various occupational health risks from nanoparticle exposure as well as their controls. It has been crucially important to understand the mode of interaction of nanomaterials (native or modified states) with biological systems so that complete assurance can be made before release of nanomaterials into the commercial markets. Further, it has been very important to identify potential exposure routes of nanomaterials (like inhalation, ingestion, or subcutaneous adsorption). The National Nanotechnology Initiative (NNI) has started various research programs to understand the social, ethical, health, and environmental implications of nanotechnology through efforts of various groups like the National Science Technology Council's Subcommittee on Nanoscale Science, Engineering and Technology (NSET). The NNI is also promoting inter- and multidisciplinary research on effective remediation, risk characterization, and communication as well as risk mitigation. The synthesis of nanomaterials involves various substances/mixtures over varying time periods with varying intensity levels; thus a substance-by-substance risk assessment approach would not be as effective [52]. In the case of modifying nanomaterials with enzymes, it is crucially important to study the overall effect of complete hybrid structures (enzyme and nanomaterials) on biological systems. Scientists are required to update their risk assessment methodologies through a multidisciplinary approach so that accurate results can be obtained. This is the best time to look for various risk assessment methodologies, as now we have almost all classes of nanomaterials that have been released for various applications in addition to their modifications. Further, there should be increased levels of cooperation between industries, public interest groups, and government parties in finding economically viable solutions while still protecting the environment and health. Further, policymakers are required to start thinking about voluntary agreements with industry on the responsible use of nanotechnology and push the development of more models that bring together universities, NGOs, and industry to develop principles and best practices. These collective efforts would have fruitful results in making nanotechnology a safe technology for environment and health, allowing it to become a long-term sustainable technology with excellent sustainable ESI scores.

5.5 Future aspects

International trade and capital flows are being strongly supported by combo-technology through long-term economic growth and development. Technological success depends on its supportive trade and investment policies as well as environmental and social policies. The Organization for Economic Cooperation and Development (OECD) countries have been striving for rapid effectual coherence between trades, investment, environmental, and social policies both at their domestic and international levels. The most concrete steps that should be taken for the broader goal of sustainable development via technological developments are:

- Reforming domestic policies leading to elimination of aspects related to trade distortion and are environmentally damaging.
- Timely assessment of environmental and social impacts of trade and investment liberalization.
- Trade and investment disciplines should be coherent with environmental and social policies.
- Both environmental and social codes of conduct should be effectively practiced in both the private and public sector.

- Government should provide supportive regulatory and institutional frameworks for private sector activity in addition to universities and institutes.
- Active cooperation between multinational enterprises and corporate governance with minimal bribery.
- Increasing market access for developing countries through economic liberalization.
- Reviewing economic and environmental policies from the perspective of the goal of poverty reduction.
- Active cooperation for international development leading to increase in GDP by maximum percent.
- Sustainable development of the global economy through technological success could be well established through policy and institutional frameworks, leading to active capital flows with minimal adverse effects on environmental and social issue impacts.
- Development at the regional and global levels can be encouraged when technology directly relates to local benefits of ordinary people with minimal adverse effects on environment and health.

References

[1] Anadon LD, Chan G, Harley AG, Matus K, Moon S, Murthy SL, Clark WC. Proc Natl Acad Sci U S A 2016;113:9682–90.
[2] Nabyonga-Orem J. BMJ Glob Health 2017;2:e000433. https://doi.org/10.1136/bmjgh-2017-000433.
[3] Haines A. Med Confl Surviv 2001;17:56–62.
[4] Thimbleby H. J Public Health Res 2013;2:e28. https://doi.org/10.4081/jphr.2013.e28.
[5] Lahr H, Mina A. Financ Manag 2014;43:291–325.
[6] Harrison J, Miller K, McNeely J. In: McNeely J, Miller K, editors. National Parks, conservation and development: the role of protected areas in sustaining society. Washington, DC: Smithsonian Institution; 1984. p. 24–33.
[7] Ridgway EM, Lawrence MA, Woods J. Front Nutr 2015;2:29. https://doi.org/10.3389/fnut.2015.00029.
[8] Elbakidze M, Hahn T, Mauerhofer V, Angelstam P, Axelsson R. Ambio 2013;42:174–87.
[9] Williams C, Blaiklock A. Int J Health Policy Manag 2016;5:387–90.
[10] Banerjee A, Halvorsen KE, Eastmond-Spencer A, Sweitz SR. Environ Manag 2017;59:912–23.
[11] Thorlakson T, de Zegher JF, Lambin EF. Proc Natl Acad Sci U S A 2018;115:2072–7.
[12] McVeigh J, MacLachlan M, Gilmore B, McClean C, Eide AH, Mannan H, Geiser P, Duttine A, Mji G, McAuliffe E, Sprunt B, Amin M, Normand C. Glob Health 2016;12:49. https://doi.org/10.1186/s12992-016-0182-8.
[13] Riedel A. Healthcare (Basel) 2016;4:2. https://doi.org/10.3390/healthcare4010002.
[14] Molfenter T, Ford JH, Bhattacharya A. Int J Inf Syst Chang Manag 2011;5:22–35.
[15] Nyström ME, Strehlenert H, Hansson J, Hasson H. BMC Health Serv Res 2014;14:401. https://doi.org/10.1186/1472-6963-14-401.

[16] Anand A, Roy N. Front Public Health 2016;4:87. https://doi.org/10.3389/fpubh.2016.00087.
[17] Stefanini A. World Health Forum 1995;16:42–6.
[18] Peschel AO, Grebitus C, Steiner B, Veeman M. Appetite 2016;106:78–91.
[19] Fields S. Environ Health Perspect 2002;110:A142–5.
[20] Obeng-quaidoo I, Gikonyo W. Afr Media Rev 1995;9:70–93.
[21] Sucala M, Nilsen W, Muench F. Transl Behav Med 2017;7:854–60.
[22] Rana S, Kalaichelvan PT. ISRN Toxicol 2013;2013:574648. https://doi.org/10.1155/2013/574648.
[23] Ray PC, Yu H, Fu PP. J Environ Sci Health C Environ Carcinog Ecotoxicol Rev 2009;27:1–35.
[24] Sharifi S, Behzadi S, Laurent S, Forrest ML, Stroeve P, Mahmoudi M. Chem Soc Rev 2012;41:2323–43.
[25] Gatoo MA, Naseem S, Arfat MY, Dar AM, Qasim K, Zubair S. Biomed Res Int 2014;2014:498420. https://doi.org/10.1155/2014/498420.
[26] Gurung N, Ray S, Bose S, Rai V. Biomed Res Int 2013;2013:329121. https://doi.org/10.1155/2013/329121.
[27] Bolibok P, Wiśniewski M, Roszek K, Terzyk AP. Naturwissenschaften 2017;104:36. https://doi.org/10.1007/s00114-017-1459-3.
[28] Pavlovskaia E. Environ Sci Eur 2014;26:17. https://doi.org/10.1186/s12302-014-0017-2.
[29] Bradshaw CJA, Giam X, Sodhi NS. PLoS ONE 2010;5:e10440. https://doi.org/10.1371/journal.pone.0010440.
[30] Cracolici MF, Cuffaro M, Nijkamp P. Soc Indic Res 2010;95:339–56.
[31] Gallego-Álvarez I, Vicente-Galindo MP, Galindo-Villardón MP, Rodríguez-Rosa M. Sustainability 2014;6:7807–32.
[32] Burger JR, Allen CD, Brown JH, Burnside WR, Davidson AD, Fristoe TS, Hamilton MJ, Mercado-Silva N, Nekola JC, Okie JG, Zuo W. PLoS Biol 2012;10:e1001345. https://doi.org/10.1371/journal.pbio.1001345.
[33] Lira-Noriega A, Soberón J. Ambio 2015;44:391–400.
[34] Quinn MM, Kriebel D, Geiser K, Moure-Eraso R. Am J Ind Med 1998;34:297–304.
[35] Mayer AL. Environ Int 2008;34:277–91.
[36] de Zea Bermudez V, Leroux F, Rabu P, Taubert A. Beilstein J Nanotechnol 2017;8:861–2.
[37] Vance ME, Kuiken T, Vejerano EP, McGinnis SP, Hochella MF, Rejeski D, Hull MS. Beilstein J Nanotechnol 2015;6:1769–80.
[38] Jakovljevic MB. Front Public Health 2015;3:135. https://doi.org/10.3389/fpubh.2015.00135.
[39] Clark WC, Dickson NM. Proc Natl Acad Sci U S A 2003;100:8059–61.
[40] Zhang F. Front Chem 2017;5:80. https://doi.org/10.3389/fchem.2017.00080.
[41] Salamanca-Buentello F, Persad DL, Court EB, Martin DK, Daar AS, Singer PA. PLoS Med 2005;2:e97. https://doi.org/10.1371/journal.pmed.0020097.
[42] Iavicoli I, Leso V, Ricciardi W, Hodson LL, Hoover MD. Environ Health 2014;13:78. https://doi.org/10.1186/1476-069X-13-78.
[43] Schulte PA, Salamanca-Buentello F. Environ Health Perspect 2007;115:5–12.
[44] Hobson DW. Wiley Interdiscip Rev Nanomed Nanobiotechnol 2009;1:189–202.
[45] Lioy PJ, Nazarenko Y, Han TW, Lioy MJ, Mainelis G. Int J Occup Environ Health 2010;16:378–87.
[46] Hood E. Environ Health Perspect 2004;112:A740–9.
[47] Singh PK, Jairath G, Ahlawat SS. J Food Sci Technol 2016;53:1739–49.

[48] Choi S, Kim S, Bae YJ, Park JW, Jung J. Environ Health Toxicol 2015;30:e2015003. https://doi.org/10.5620/eht.e2015003.
[49] Butler PJ, Dey KK, Sen A. Cell Mol Bioeng 2015;8:106–18.
[50] Sharma A, Madhunapantula SV, Robertson GP. Expert Opin Drug Metab Toxicol 2012;8:47–69.
[51] Chen M, Zeng G, Xu P, Lai C, Tang L. Trends Biochem Sci 2017;42:914–30.
[52] Laux P, Tentschert J, Riebeling C, Braeuning A, Creutzenberg O, Epp A, Fessard V, Haas K-H, Haase A, Hund-Rinke K, Jakubowski N, Kearns P, Lampen A, Rauscher H, Schoonjans R, Störmer A, Thielmann A, Mühle U, Luch A. Arch Toxicol 2018;92:121–41.

Index

Note: Page numbers followed by *f* indicate figures and *t* indicate tables.

D

E

F

G

H

I

K

L

W

Z